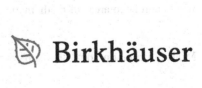

Advances in Mathematical Fluid Mechanics

Lecture Notes in Mathematical Fluid Mechanics

Editor-in-Chief
Giovanni P Galdi
University of Pittsburgh, Pittsburgh, PA, USA

Series Editors
Didier Bresch
Université Savoie-Mont Blanc, Le Bourget du Lac, France

Volker John
Weierstrass Institute, Berlin, Germany

Matthias Hieber
Technische Universität Darmstadt, Darmstadt, Germany

Igor Kukavica
University of Southern California, Los Angles, CA, USA

James Robinson
University of Warwick, Coventry, UK

Yoshihiro Shibata
Waseda University, Tokyo, Japan

Lecture Notes in Mathematical Fluid Mechanics as a subseries of "Advances in Mathematical Fluid Mechanics" is a forum for the publication of high quality monothematic work as well lectures on a new field or presentations of a new angle on the mathematical theory of fluid mechanics, with special regards to the Navier-Stokes equations and other significant viscous and inviscid fluid models.

In particular, mathematical aspects of computational methods and of applications to science and engineering are welcome as an important part of the theory as well as works in related areas of mathematics that have a direct bearing on fluid mechanics.

More information about this subseries at http://www.springer.com/series/15480

Wojciech S. Ożański

The Partial Regularity Theory of Caffarelli, Kohn, and Nirenberg and its Sharpness

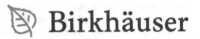 Birkhäuser

Wojciech S. Ożański
Department of Mathematics
University of Southern California
Los Angeles, CA, USA

ISSN 2297-0320   ISSN 2297-0339   (electronic)
Advances in Mathematical Fluid Mechanics
ISSN 2510-1374   ISSN 2510-1382   (electronic)
Lecture Notes in Mathematical Fluid Mechanics
ISBN 978-3-030-26660-8   ISBN 978-3-030-26661-5   (eBook)
https://doi.org/10.1007/978-3-030-26661-5

Mathematics Subject Classification (2010): 35Q30, 76D05, 76D03

This book is published under the imprint Birkhäuser, www.birkhauser-science.com by the registered
company Springer Nature Switzerland AG
The registered company address is: Gewerbestrasse 11, 6330 Cham, Switzerland

Contents

Chapter 1
Introduction

Abstract We present the context of the partial regularity theory of the three-dimensional incompressible Navier-Stokes equations due to Caffarelli, Kohn and Nirenberg.

We present the context of the partial regularity theory of the three-dimensional incompressible Navier-Stokes equations due to Caffarelli, Kohn and Nirenberg. We discuss the concept of weak solutions of the Navier-Stokes inequality and introduce the constructions of such solutions (due to Scheffer) which show the sharpness of the Caffarelli–Kohn–Nirenberg theorem.

The Navier–Stokes equations,

$$\partial_t u - \nu \Delta u + (u \cdot \nabla)u + \nabla p = 0,$$
$$\operatorname{div} u = 0, \tag{1.1}$$

where u denotes the velocity of a fluid, p the scalar pressure and $\nu > 0$ the viscosity, comprise a fundamental model for viscous, incompressible flows. The equations are usually supplemented with an initial condition $u(0) = u_0$, where a divergence-free vector field u_0 is given.

The fundamental mathematical theory of the Navier–Stokes equations goes back to the pioneering work of Leray (1934) (see Ożański & Pooley, 2018 for a comprehensive modern review of this paper), who used an elegant Picard iteration scheme to prove existence and uniqueness of local-in-time strong solutions on the whole space $\mathbb{R}^3$.

Definition 1.1 (*Strong solution of the NSE*) We say that u is a *strong solution* of the Navier–Stokes equations (1.1) on $\Omega \times (0, T)$ with the initial condition $u_0 \in V$ if

(i) (regularity) $u \in L^\infty((0, T); V) \cap L^2((0, T); H^2(\Omega))$,
(ii) (the equation) u satisfies

$$\int_0^T \int_\Omega (-u \cdot \partial_t \phi + \nu \nabla u : \nabla \phi + (u \otimes u) : \nabla \phi) = \int_\Omega u_0 \cdot \phi(0)$$

© Springer Nature Switzerland AG 2019
W. S. Ożański, *The Partial Regularity Theory of Caffarelli, Kohn, and Nirenberg and its Sharpness*, Advances in Mathematical Fluid Mechanics, https://doi.org/10.1007/978-3-030-26661-5_1

for all divergence-free $\phi \in C_0^\infty(\Omega \times [0, T); \mathbb{R}^3)$

Here

$$V := \{v \in H^1(\mathbb{R}^3; \mathbb{R}^3) : \operatorname{div} v = 0\}$$

if $\Omega = \mathbb{R}^3$, where H^1 denotes the Sobolev space of functions $v \in L^2$ such that $\nabla v \in L^2$. In the case $\Omega \neq \mathbb{R}^3$ we set V to be the closed subspace of $\{v \in H^1(\Omega; \mathbb{R}^3) :$ div $v = 0\}$ which takes into account appropriate boundary conditions. For example, if $\Omega \subset \mathbb{R}^3$ is a bounded, smooth domain then $V := \{v \in H^1(\Omega; \mathbb{R}^3) : \operatorname{div} v = 0, v|_{\partial\Omega} = 0\}$, and if $\Omega = \mathbb{T}^3$ (the three-dimensional flat torus) then $V := \{v \in H^1(\mathbb{T}^3; \mathbb{R}^3) : \operatorname{div} v = 0, \int_{\mathbb{T}^3} v = 0\}$.

We note that the above definition does not include the pressure function p, which can be recovered from the equation $-\Delta p = \partial_{ij}(u_i u_j)$ (which in turn can be derived by calculating the divergence of (1.1); see Chap. 5 of Robinson, Rodrigo & Sadowski, 2016 for details). It can be shown in the case $\Omega = \mathbb{R}^3$ that p is given uniquely by

$$p = \sum_{i,j=1}^{3} \partial_{ij}\Psi * (u_i u_j) \tag{1.2}$$

at every time (see (6.47) in Ożański & Pooley, 2018 for details), where $\Psi(x) := (4\pi|x|)^{-1}$ is the fundamental solution of the Laplace equation.

In addition to the notion of strong solutions, one often considers the so-called *Leray–Hopf weak solutions*.

Definition 1.2 (*Leray–Hopf weak solutions of the NSE*) A vector field $u \in L^2$ $((0, T); V)$ is a *Leray–Hopf weak solution* of the Navier–Stokes equations (1.1) on $\Omega \times (0, T)$ with the divergence-free initial condition $u_0 \in L^2(\Omega)$ if it satisfies condition (ii) of Definition 1.1 and the energy inequality

$$\|u(t)\|_{L^2}^2 + 2\nu \int_s^t \|\nabla u(\tau)\|_{L^2}^2 \mathrm{d}\tau \leq \|u(s)\|_{L^2}^2 \tag{1.3}$$

for almost every $s \geq 0$ and every $t > s$.

Solutions defined above are named after Leray (1934) and Hopf (1951), who were the first to construct such solutions that are global in time in the case of the whole space $\mathbb{R}^3$ (Leray) and a bounded, smooth domain $\Omega \subset \mathbb{R}^3$ (Hopf). They can be thought of as weak continuations of a strong solution, due to a certain weak-strong uniqueness result, see, for example, Theorem 6.10 in Robinson et al. (2016) or Lemma 6.39 in Ożański and Pooley (2018). We refer the reader to Robinson et al. (2016) for a more comprehensive introduction to the theory of existence and uniqueness of solutions as well as other topic related to the Navier–Stokes equations (1.1).

Although the fundamental question of global-in-time existence and uniqueness of strong solutions remains unresolved (as does the question of uniqueness of Leray-Hopf weak solutions; however, see Buckmaster and Vicol (2019) for nonuniqueness of (non-Leray–Hopf) weak solutions), many significant results contributed to the theory of the Navier–Stokes equations during the second half of the twentieth century.

The Caffarelli–Kohn–Nirenberg Theorem

One such contribution is the partial regularity theory introduced by Scheffer (1976a, b, 1977, 1978, 1980) and subsequently developed by Caffarelli, Kohn and Nirenberg (1982); see also Lin (1998), Ladyzhenskaya and Seregin (1999), Vasseur (2007) and Kukavica (2009b) for alternative approaches.

This theory is concerned with local behaviour of weak solutions, which gives rise to the notion of a "suitable" weak solution.

Definition 1.3 (*Suitable weak solution of the NSE*) A pair (u, p) is a *suitable weak solution* of the Navier–Stokes equations with viscosity $\nu > 0$ on an open set $U \times (a, b)$ if

(i) (regularity of u and p) $u \in L^\infty((a, b); L^2(U))$, $\nabla u \in L^2(U \times (a, b))$, $u(t)$ is divergence-free for almost every $t \in (a, b)$, and $p \in L^{3/2}_{loc}(U \times (a, b))$,

(ii) (relation between u and p) the equation $-\Delta p = \sum_{i,j=1}^{3} \partial_i \partial_j (u_i u_j)$ holds in the sense of distributions in U for almost every $t \in (a, b)$,

(iii) (the local energy inequality) the inequality

$$\int_U |u(t)|^2 \phi(t) + 2\nu \int_a^t \int_U |\nabla u|^2 \phi$$
$$\leq \int_a^t \int_U \left(|u|^2 (\partial_t \phi + \nu \Delta \phi) + (|u|^2 + 2p)(u \cdot \nabla)\phi \right) \tag{1.4}$$

is valid for every $\phi \in C_0^\infty(U \times (a, b); [0, \infty))$ and $t \in (a, b)$.

(iv) (the equation) the Navier–Stokes equation (1.1) holds in the sense of distributions on $U \times (a, b)$, that is

$$\int_a^b \int_U (u \cdot (\partial_t \phi + \nu \Delta \phi) - \phi \cdot ((u \cdot \nabla)u) + p \operatorname{div} \phi) = 0 \tag{1.5}$$

for all $\phi \in C_0^\infty(U \times (a, b); \mathbb{R}^3)$.

Note that, in contrast to (ii) in Definition 1.1, (1.5) must hold for all ϕ (i.e. not only for divergence-free ϕ), and in particular the pressure term in (1.5) does not vanish. Note also that the regularity assumptions on u (i.e. boundedness in time of the L^2 norm and the space-time L^2 integrability of the gradient) is the same as the regularity of Leray-Hopf weak solutions that can be deduced from the energy inequality (1.3). Using the regularity of u from (i) and the Lebesgue interpolation one obtains that

$$u \in L^{10/3}(U \times (a, b)), \tag{1.6}$$

see Lemma 3.5 in Robinson et al. (2016), for example. Thus (i) implies that all terms on the right-hand side of the local energy inequality (1.4) are well defined. Moreover, in the case when $U = \mathbb{R}^3$ and p is given by (1.2) (ii) is satisfied and

$$p \in L^{5/3}(\mathbb{R}^3 \times (a, b)), \tag{1.7}$$

which can be deduced from (1.6) using the Calderón-Zygmund inequality (see (2.39) below). In other words in this case the regularity of p required in (i) (i.e. $p \in L_{loc}^{3/2}(U \times (a, b)))$ follows from the regularity of u.

An important difference between suitable weak solutions and Leray-Hopf weak solutions is that the former is a distributional solution of the NSE, while the latter is a solution of the initial value problem (i.e. (1.1) with $u(0) = u_0$). In addition to this, suitable weak solutions satisfy the local energy inequality (1.4), which is an interior regularity assumption not included in the definition of Leray-Hopf weak solutions (Definition 1.2). However, given divergence-free initial data $u_0 \in L^2(\mathbb{R}^3)$, there exist Leray-Hopf weak solutions that are suitable, as was proved by Scheffer (1977) (and by Caffarelli et al. (1982) in the case of a bounded domain). In fact, the Leray–Hopf weak solutions on $\mathbb{R}^3$ constructed by Leray (1934) are suitable, which can be deduced from Theorem 2.1 in Biryuk, Craig and Ibrahim (2007).

The central result of the partial regularity theory is the following theorem, which was proved by Caffarelli et al. (1982).

Theorem 1.4 (Partial regularity of the Navier–Stokes equations) *There exist $\varepsilon_0, \varepsilon_1 > 0$ with the following properties. Let (u, p) be a suitable weak solution of the Navier–Stokes equations on $\mathbb{R}^3 \times (0, \infty)$, and let*

$$Q_r = Q_r(z) := \{(x, t): |x - y| < r, \, t \in (s - r^2, s]\} \subset \mathbb{R}^3 \times (0, \infty),$$

where $z = (y, s)$, denote a cylinder in space-time. Then

(i) if

$$\frac{1}{r^2} \int_{Q_r} \left(|u|^3 + |p|^{3/2}\right) \le \varepsilon_0 \tag{1.8}$$

then $u \in L^\infty(Q_{r/2})$;

(ii) if

$$\limsup_{r \to 0} \frac{1}{r} \int_{Q_r} |\nabla u|^2 \le \varepsilon_1 \tag{1.9}$$

then $u \in L^\infty(Q_\rho)$ for some $\rho > 0$.

We note that alternative approaches to partial regularity have been developed by Lin (1998), Ladyzhenskaya and Seregin (1999), Vasseur (2007) and Kukavica (2009b). In fact, the results of Lin (1998), Ladyzhenskaya and Seregin (1999) are a little different, as instead of local boundedness (as in the theorem above) they show a stronger property, namely local Hölder continuity (in space-time).

In short, the above theorem provides sufficient conditions on the local (in space-time) behaviour of suitable weak solutions that guarantee boundedness. The partial regularity theorem is also a key ingredient in the $L_{3,\infty}$ regularity criterion for the three-dimensional Navier–Stokes equations (see Escauriaza, Seregin & Šverák, 2003) and in the uniqueness of Lagrangian trajectories for suitable weak solutions (Robinson & Sadowski, 2009); similar ideas have also been used for other models, such as the surface growth model

$$\partial_t u + u_{xxxx} + \partial_{xx} u_x^2 = 0$$

(Ożański & Robinson, 2019), which is a one-dimensional model of the Navier–Stokes equations (Blömker & Romito 2009, 2012).

A remarkable feature of this partial regularity result is that the quantities involved (namely $|u|^3$, $|p|^{3/2}$, $|\nabla u|^2$) are globally (in space-time) integrable for any Leray-Hopf weak solution. This leads to a corollary of Theorem 1.4 which provides upper bounds on the dimension of a putative singular set.

The Singular Set of a Suitable Weak Solution of the Navier–Stokes Equations

Theorem 1.4 implies that, given a suitable weak solution (u, p), there cannot be "too many" singular points. This point can be made precise by considering the *singular set*, i.e.

$$S := \{(x, t) \in U \times (a, b) \colon u \text{ is unbounded in any neighbourhood of } (x, t)\}. \tag{1.10}$$

In other words, $(x, t) \notin S$ if and only if u is bounded in a neighbourhood of (x, t). In fact, it then follows from a local Serrin condition that u is smooth in the spatial variables in such a neighbourhood (see Theorem 13.4 in Robinson et al. 2016 for a proof). Theorem 1.4 lets us estimate the dimension of S.

Corollary 1.5 *Let (u, p) be a suitable weak solution of the Navier–Stokes equations on $U \times (a, b)$. Then*

$$d_H(S) \leq 1. \tag{1.11}$$

Moreover, if $p \in L_{loc}^{5/3}(U \times (a, b))$ then for every compact $K \subset U \times (a, b)$

$$d_B(S \cap K) \leq 5/3.$$

Here d_B denotes the (upper) box-counting dimension and d_H denotes the Hausdorff dimension. Namely, given a compact set K

$$d_B(K) := \limsup_{r \to 0} \frac{M(K, r)}{-\log r}, \tag{1.12}$$

where $M(K, r)$ stands for the maximal number of r-balls with centres in K (or, equivalently, for the minimal number of r-balls required to cover K, see Exercise 3.1 in Robinson (2011) for a proof of this equivalence and a number of other equivalent definitions) and

$$d_H(K) := \inf\{s \geq 0 \colon \mathcal{H}^s(K) = 0\}, \tag{1.13}$$

where

$$\mathcal{H}^s(K) := \lim_{\delta \to 0^+} \inf \left\{ \sum_k (\operatorname{diam}(B_k))^s \colon \operatorname{diam}(B_k) \leq \delta \text{ and } \{B_k\} \text{ covers } K \right\} \tag{1.14}$$

denotes the s-dimensional Hausdorff measure.

We point out that Theorem 1.4 actually implies an estimate that is stronger than (1.11). Namely $\mathcal{P}^1(S) = 0$ (see Lemma 2.8 for details), where $\mathcal{P}^1$ denotes the one-dimensional *parabolic* Hausdorff measure (that is (1.14) with balls B_k replaced by cylinders Q_k).

The point of considering the intersection $S \cap K$ is to separate S from the boundary of the domain $U \times (a, b)$, which is a technical matter related to the definition of the box-counting dimension, which we discuss below.

The Hausdorff Dimension and the Box-Counting Dimension

An interesting fact about the two notions of dimension (i.e. the Hausdorff dimension and the Minkowski dimension) is that they both extend the "usual" notion of dimension to non-integer values. Namely, if $I \subset \mathbb{R}^3$ is a line segment, $L \subset \mathbb{R}^3$ is a two-dimensional surface (e.g. a unit sphere) and $Q \subset \mathbb{R}^3$ is a three-dimensional body (e.g. a unit cube or ball) then

$$d_H(I) = d_B(I) = 1, \quad d_H(L) = d_B(L) = 2, \quad d_H(Q) = d_B(Q) = 3,$$

see Lemma 3.3 (vi) in Robinson (2011) and Sect. 3.2 in Falconer (2014).

There exist many well-known fractal sets K for which $d_H(K) = d_B(K)$ (for example Cantor sets) and also many fractal sets K for which $d_H(K) < d_B(K)$; for example the set $K := \{n^\alpha \colon n \in \mathbb{N}\} \subset \mathbb{R}$ satisfies $d_B(K) = (\alpha + 1)^{-1}$, while $d_H(K) = 0$, since the Hausdorff dimension vanishes for countable sets (which can be seen directly from the definition (1.13)).

Given a compact set $K \subset \mathbb{R}^3$, both $d_H(K)$ and $d_B(K)$ measure, in a sense, the complexity of K. As can be deduced from the above definitions (1.12), (1.13), the box-counting dimension is determined by counting the minimal number of r-balls required to cover K as $r \to 0^+$, while the Hausdorff dimension is concerned with the sum of diameters (taken to some power) of elements of any cover. It is this difference that gives the general inequality

$$d_H(K) \leq d_B(K) \qquad \text{for any compact } K, \tag{1.15}$$

and that gives the difference in the two bounds in Corollary 1.5. Indeed, if $(x, t) \in S$ then $Q_r(x, t)$ does not satisfy (1.8) **for any** $r > 0$, as long as $Q_r(x, t) \Subset U \times (a, b)$ (which is the reason why we consider the intersection $S \cap K$, instead of S). On the other hand (1.9) implies merely that $\rho^{-1} \int_{Q_\rho(x,t)} |\nabla u|^2 > \varepsilon_1$ for **some** sufficiently small $\rho > 0$ (which may vary depending on (x, t)). In other words, part (i) of Theorem 1.4 allows us to construct covers of S of any given radius (and therefore deduce a bound on $d_B(S)$), while the condition (1.9) does not allow us to guarantee the same radius of the sets in a cover, and consequently it implies merely the bound on $d_H(S)$. This point is demonstrated in more detail in the proof of Corollary 1.5 in Section 2.4.

Interestingly, the bound on the box-counting dimension in Corollary 1.5 (which was first proved by Robinson & Sadowski, 2009) has been improved. First Kukavica (2009a) showed that $d_B(S \cap K) \le 135/82 (\approx 1.65)$ for every compact K. This bound was later refined by Kukavica and Pei (2012), Koh and Yang (2016) down to the most recent bound $d_B(S \cap K) \le 2400/1903 (\approx 1.261)$ obtained by He, Wang and Zhou (2017). As for the Hausdorff dimension, the bound $d_H(S) \le 1$ has not been improved (although we point out an intriguing result by Choe and Lewis (2000), which provides conditions regarding merely the boundedness (rather than smallness) of the quantities in (1.8), (1.9) that guarantee that $\mathcal{P}^{1-\varepsilon}(S) = 0$ for some $\varepsilon > 0$). In what follows we present a result suggesting that such an improvement might be very difficult.

The Navier–Stokes Inequality

Another remarkable property of the partial regularity Theorem 1.4 is that it applies to a much wider family of vector fields than merely the suitable weak solutions. To be precise, from the properties (i)–(iv) of suitable weak solutions only (i)–(iii) are used in the proof of Theorem 1.4 which is due to to Caffarelli et al. (1982). In other words the property that (u, p) solve the Navier–Stokes equations (in the sense of distributions) is irrelevant to the claim of Theorem 1.4. We note, however, that this is not the case in the alternative proofs of the theorem, due to Lin (1998), Ladyzhenskaya and Seregin (1999), Vasseur (2007) and Kukavica (2009b). This observation gives rise to the notion of a weak solution of the Navier–Stokes inequality.

Definition 1.6 (*Weak solutions to the Navier–Stokes inequality*) A pair (u, p) is a *weak solution of the Navier–Stokes inequality* with viscosity $\nu > 0$ on $\mathbb{R}^3 \times (0, \infty)$ if it satisfies conditions (i)–(iii) of Definition 1.3.

We use the name *Navier–Stokes inequality* since the local energy inequality (1.4),

$$\int_U |u(t)|^2 \phi(t) + 2\nu \int_a^t \int_U |\nabla u|^2 \phi \le \int_a^t \int_U \left(|u|^2(\partial_t \phi + \nu \Delta \phi) + (|u|^2 + 2p)(u \cdot \nabla)\phi \right),$$

where $\phi \in C_0^\infty(U \times (a, b); [0, \infty))$, is a weak form of the inequality

$$u \cdot (\partial_t u - \nu \Delta u + (u \cdot \nabla)u + \nabla p) \le 0. \tag{1.16}$$

Indeed, assuming that (u, p) is smooth (in both space and time) we can rewrite (1.16) in the form

$$\frac{1}{2}\partial_t |u|^2 - \frac{\nu}{2}\Delta |u|^2 + \nu|\nabla u|^2 + u \cdot \nabla\left(\frac{1}{2}|u|^2 + p\right) \le 0,$$

where we used the calculus identity $u \cdot \Delta u = \Delta(|u|^2/2) - |\nabla u|^2$. Multiplication by 2φ and integration by parts gives (1.4).

Furthermore, setting

$$f := \partial_t u - \nu\Delta u + (u \cdot \nabla)u + \nabla p, \qquad (1.17)$$

one can (formally) think of the Navier–Stokes inequality (1.16) as the inhomogeneous Navier–Stokes equations with forcing f,

$$\partial_t u - \nu\Delta u + (u \cdot \nabla)u + \nabla p = f,$$

where f acts against the direction of the flow u, that is $f \cdot u \le 0$.

Observe also that the energy inequality (1.3) can be derived (formally) by integrating (1.16) over $\Omega \times (s, t)$.

The "Real" Caffarelli–Kohn–Nirenberg Theorem

Having defined weak solutions to the NSI, we can now state the Caffarelli–Kohn–Nirenberg theorem in its full extent.

Theorem 1.7 (Partial regularity of weak solutions to the NSI) *Let u be a weak solution of the Navier–Stokes inequality. Then the claims of Theorem 1.4 remain true. In particular*

$$d_H(S) \le 1,$$

where S is the singular set of u.

The aim of Chap. 2 is to provide a self-contained proof of the above theorem. We note that the main iteration scheme in our proof of part (i) of Theorem 1.7 does not involve the pressure function (see Sect. 2.2), which is a significant simplification of the classical proof.

The Sharpness of the CKN Theorem

Even more remarkably, the claim of the above theorem is sharp, as was shown by Scheffer (1987).

Theorem 1.8 (Scheffer's counterexample) *Given $\xi \in (0, 1)$ there exists a weak solution u to the Navier–Stokes inequality on $\mathbb{R}^3 \times (0, \infty)$ with*

$$\xi \le d_H(S) \le 1.$$

Furthermore $u(t) \in C_0^\infty(\mathbb{R}^3)$ with $\operatorname{supp} u(t) \subset G$ for all $t > 0$, where $G \subset \mathbb{R}^3$ is compact.

The aim of Chaps. 3 and 4 is to present an approachable proof of Theorem 1.8. We will show (in Chap. 4) how the theorem above can be proved by extending the earlier result by Scheffer (1985), which we now state.

Theorem 1.9 (Weak solution of NSI with point singularity) *There exist $\nu_0 > 0$ and a function $\mathfrak{u} \colon \mathbb{R}^3 \times [0, \infty) \to \mathbb{R}^3$ that is a weak solution of the Navier–Stokes inequality for all $\nu \in [0, \nu_0]$ such that $\mathfrak{u}(t) \in C_0^\infty$, $\operatorname{supp} \mathfrak{u}(t) \subset G$ for all t for some compact set $G \Subset \mathbb{R}$ (independent of t). Moreover $\mathfrak{u}$ is unbounded in every neighbourhood of (x_0, T_0), for some $x_0 \in G$, $T_0 > 0$.*

It is clear, using an appropriate rescaling, that the statement of the above theorem is equivalent to the one where $\nu_0 = 1$ and $(x_0, T_0) = (0, 1)$. Indeed, if $\mathfrak{u}$ is the velocity field given by the theorem then $\sqrt{T_0/\nu_0}\mathfrak{u}(x_0 + \sqrt{T_0\nu_0}x, T_0 t)$ satisfies Theorem 1.9 with $\nu_0 = 1$, $(x_0, T_0) = (0, 1)$. We articulate that the vector field $\mathfrak{u}$ given by Theorem 1.9 satisfies the Navier–Stokes inequality *for all values of viscosity ν* between 0 and ν_0.

In Chap. 3 we present a proof of Theorem 1.9 that is more succinct and intuitive than the original. As a part of the simplification process we introduce the notion of a *structure* on an open subset of the upper half-plane (see Definition 3.4), which allows one to construct a compactly supported, divergence-free vector field u in $\mathbb{R}^3$ with prescribed absolute value $|u|$ and with a number of other useful properties (see Sect. 3.2.4 and Lemma 3.2). Moreover, we point out the key concepts used in the construction of the blow-up. Namely, we introduce the notion of the *pressure interaction function* (corresponding to a given subset of the half-plane and its structure, see Sect. 3.2.6), which articulates a certain nonlocal property of the pressure function (see Lemma 3.6). In particular we explain how can one engineer a weak solution of the NSI with a blow-up via the term

$$u \cdot \nabla p$$

in the NSI (1.16). This is remarkable since this term vanishes after integration over $\mathbb{R}^3$, and thus, in particular, its behaviour is not captured by the energy inequality (1.3).

What is more, we formalise the concept of the *geometric arrangement* (see Sect. 3.3), that is a certain configuration of subsets of the upper half-plane (and their structures) which, in a sense, "magnifies" the pressure interaction. We also expose some other concepts used in the proof, such as an analysis of rescalings of vector fields and some ideas related to dealing with the nonlocal character of the pressure function. In addition to these simplifications, we point out how Theorem 1.8 is obtained as a straightforward extension of Theorem 1.9.

Furthermore, we improve Theorem 1.8 in the case $\nu_0 = 0$ to construct weak solutions to the "Euler inequality",

$$u \cdot (\partial_t u + (u \cdot \nabla)u + \nabla p) \leq 0,$$

which, except for a finite-time blow-up on a Cantor set, also satisfy the "almost equality"

$$-\vartheta \le u \cdot (\partial_t u + (u \cdot \nabla)u + \nabla p) \le 0 \qquad (1.18)$$

for any preassigned $\vartheta > 0$, in the sense that we now make precise. We will divide the time interval $(0, \infty)$ into countably many disjoint open intervals $\{I_k\}$ such that $\bigcup \overline{I_k} = [0, \infty)$, where $\overline{I_k}$ denotes the closure of I_k. We will be concerned with vector fields u that are smooth on $\mathbb{R}^3 \times I_k$ for each k and that $u|_{I_{k+1}}$ and $u|_{I_k}$ can be combined (one after another) to yield a weak solution of the NSI. Such a switching procedure (which will become clear in Chap. 3) does not allow us to define the time derivative at the switching time and for this reason we will understand (1.18) in the sense that

$$-\vartheta \le u \cdot (\partial_t u + (u \cdot \nabla)u + \nabla p) \le 0 \quad \text{everywhere in } \mathbb{R}^3 \times I_k \text{ for every } k.$$
$$(1.19)$$

The definition of the intervals $\{I_k\}$ will be a part of the construction of u.

In order to obtain such a result we use the construction from Chap. 3 and present a simple argument showing how the almost equality requirement (1.18) (with any ϑ) enforces $\nu = 0$; we thereby obtain the following result.

Theorem 1.10 *Given $\xi \in (0, 1)$ and $\vartheta > 0$ there exists a weak solution u to the Navier–Stokes inequality with viscosity $\nu = 0$ such that $\xi \le d_H(S) \le 1$, where S is the singular set of u, and such that (1.18) holds with $\nu = 0$ (i.e. (1.19) holds with $\nu = 0$ for some choice of the intervals $\{I_k\}$).*

In other words, there exists a divergence-free solution of the inhomogeneous Euler equation,

$$\partial_t u + (u \cdot \nabla)u + \nabla p = f,$$

with the forcing f "almost orthogonal" to the velocity field, that is $-\vartheta \le u \cdot f \le 0$, and that blows up on the Cantor set.

It is not clear how to obtain a weak solution of the Navier–Stokes inequality (with some $\nu > 0$) that blows up and satisfies the almost equality. However, one can sharpen Scheffer's constructions to obtain the following "norm inflation" result.

Theorem 1.11 (Smooth solution of NSI with norm inflation) *Given $\mathcal{N} > 0$, $\vartheta > 0$ there exist $\eta > 0$ a classical solution $u \in C^\infty(\mathbb{R}^3 \times (-\eta, 1 + \eta); \mathbb{R}^3)$ to the Navier–Stokes inequality (1.16) for all $\nu \in [0, 1]$ satisfying the almost equality*

$$\|u \cdot (\partial_t u - \nu \Delta u + (u \cdot \nabla)u + \nabla p)\|_{L^\infty} \le \vartheta, \qquad (1.20)$$

for all $\nu \in [0, 1]$, $\operatorname{supp} u(t) = G$ for all t (where $G \subset \mathbb{R}^3$ is compact), and

$$\|u(1)\|_{L^\infty} \ge \mathcal{N} \|u(0)\|_{L^\infty}.$$

The role of $\eta > 0$ in the above statement is not essential; it only articulates that, roughly speaking, nothing bad happens at times 0 and T.

We also point out that it has been recently shown that one can extend Theorem 1.8 to obtain a weak solution of the Navier–Stokes inequality such that, except for a blow-up on S with $\xi \leq d_H(S) \leq 1$, the energy $\|u(t)\|_{L^2(\mathbb{R}^3)}$ follows any prescribed, nonincreasing function, up to arbitrarily small $\varepsilon > 0$, see Ożański (2018) for details.

The structure of the manuscript is as follows.

In Chap. 2 we give a simple proof of the Caffarelli–Kohn–Nirenberg Theorem 1.7. For this purpose we first introduce the relevant scale-invariant quantities and prove a simple interpolation inequality in Sect. 2.1. We then prove part (i) of the theorem in Sect. 2.2 and part (ii) in Sect. 2.3. Each of the two parts uses a particular form of a local-in-space estimate of the pressure function p, which we prove in Sect. 2.5. The chapter is concluded with Sect. 2.4, which proves Corollary 1.5.

In Chap. 3 we prove Theorem 1.9. We start by presenting a simple sketch proof of the theorem in Sect. 3.1, which is followed by a number of observations regarding the constructed solution. The sketch is based on the existence of certain objects, which, after introducing a number of preliminary concepts in Sect. 3.2, we construct in Sect. 3.3. The construction of these objects is based on a certain "geometric arrangement", which we discuss in Sect. 3.4. At the end of the chapter, in Sect. 3.5, we prove Theorem 1.11, which is a corollary of Theorem 1.9.

In Chap. 4 we prove Theorem 1.8. After introducing some useful notation related to constructing Cantor set in Sect. 4.1 we give a sketch proof in Sect. 4.2. The sketch is based on some extensions of ideas from Chap. 3, which we present in Sects. 4.3–4.5. Finally, we prove Theorem 1.10 (which is a corollary of Theorem 1.8) in Sect. 4.6.

Acknowledgements The author would like to thank James Robinson for his enthusiasm and interest in this work as well as for reading various versions of the manuscript. His numerous comments greatly improved the quality of the text. This work arose in part from the fluid mechanics reading group organised at the University of Warwick by James Robinson and José Rodrigo. The author has been supported partially by EPSRC as part of the MASDOC DTC at the University of Warwick, Grant No. EP/HO23364/1, and partially by postdoctoral funding from ERC 616797.

Chapter 2
The Caffarelli–Kohn–Nirenberg Theorem

In this chapter, we prove the Caffarelli–Kohn–Nirenberg theorem (Theorem 1.7), where we will also assume that $\nu = 1$. Namely, we will be concerned with weak solutions of the Navier–Stokes inequality on a given open set $U \times (a, b)$, that is, according to Definition 1.6, pairs (u, p) such that $u \in L^\infty((a, b); L^2(U))$, $\nabla u \in L^2(U \times (a, b))$, $u(t)$ is divergence-free for almost every $t \in (a, b)$, $p \in L^{3/2}_{loc}(U \times (a, b))$ (**the regularity properties**), the equation $-\Delta p = \sum_{i,j=1}^3 \partial_i \partial_j (u_i u_j)$ holds in the sense of distributions in U for almost every $t \in (a, b)$ (**the relation between u and p**), and

$$\int_U |u(t)|^2 \phi(t) + 2 \int_a^t \int_U |\nabla u|^2 \phi \leq \int_a^t \int_U \left(|u|^2 (\partial_t \phi + \Delta \phi) + (|u|^2 + 2p)(u \cdot \nabla)\phi \right)$$
(2.1)

is valid for every $\phi \in C_0^\infty(U \times (a, b); [0, \infty))$ and $t \in (a, b)$ (**the local energy inequality**). We will prove that

(i) for each $Q_r \subset U \times (a, b)$ the condition

$$\frac{1}{r^2} \int_{Q_r} \left(|u|^3 + |p|^{3/2} \right) \leq \varepsilon_0$$

implies that $u \in L^\infty(Q_{r/2})$, and

(ii) if

$$\limsup_{r \to 0} \frac{1}{r} \int_{Q_r} |\nabla u|^2 \leq \varepsilon_1$$

then $u \in L^\infty(Q_\rho)$ for some $\rho > 0$.

Here $\varepsilon_0, \varepsilon_1 > 0$ are universal constants and we used the notation

$$Q_\rho = Q_\rho(y, s) = B_\rho(y) \times (s - \rho^2, s],$$
(2.2)

© Springer Nature Switzerland AG 2019
W. S. Ożański, *The Partial Regularity Theory of Caffarelli, Kohn, and Nirenberg and its Sharpness*, Advances in Mathematical Fluid Mechanics, https://doi.org/10.1007/978-3-030-26661-5_2

where $B_\rho(y) := \{x \in \mathbb{R}^3 : |x - y| < \rho\}$ denotes an open ball in $\mathbb{R}^3$. We articulate that Q_ρ is a *non-anticipating* cylinder, namely, it is based at the space time point (y, s) which lies at the upper (with respect to time) lid of the Q_ρ (rather than in the interior of Q_ρ). For simplicity, we will often write $B_\rho \equiv B_\rho(y)$.

We point out that Q_ρ includes its upper lid (i.e. $B_\rho \times \{s\}$) and so, in particular, it is not an open set. We apply such a convention due to a notational convenience. Namely, we will often consider cut-off functions χ such that, for example, $\chi = 1$ on $Q_{\rho/2}$ with $\chi \in C_0^\infty(Q_\rho; [0, \infty))$. The latter is a shorthand notation which means that χ is a restriction to Q_ρ of a nonnegative, smooth function with compact support in $B_\rho \times (s - \rho^2, s + \varepsilon)$ for some $\varepsilon > 0$.

2.1 Preliminaries

Throughout the chapter, we denote by "c" any absolute constant greater than 1 (whose value might change at each appearance). For brevity, we also write $\| \cdot \|_p \equiv \| \cdot \|_{L^p(\mathbb{R}^3)}$ as well as $\partial_i \equiv \partial_{x_i}$ and $\partial_{ij} \equiv \partial_i \partial_j$.

2.1.1 Scale Invariance

We first note that all of the quantities involved in the local regularity criteria in (i) and (ii), namely $r^{-2} \int_{Q_r} |u|^3$, $r^{-2} \int_{Q_r} |p|^{3/2}$ and $r^{-1} \int_{Q_r} |\nabla u|^2$, are scale-invariant. To be more precise, it is easy to check that the Navier–Stokes equations (and the Navier–Stokes inequality) are invariant under the scaling

$$\begin{cases} u(x, t), \\ p(x, t) \end{cases} \mapsto \begin{cases} u_\lambda(x, t) := \lambda u(\lambda x + x_0, \lambda^2 t + t_0), \\ p_\lambda(x, t) := \lambda^2 p(\lambda x + x_0, \lambda^2 t + t_0) \end{cases}$$

for any $\lambda > 0$ and $x_0 \in \mathbb{R}^3$, $t_0 \in \mathbb{R}$. By the scale invariance of each of the quantities above, say $r^{-2} \int_{Q_r} |u|^3$ (for example), we mean that

$$r^{-2} \int_{Q_r} |u|^3 = (r/\lambda)^{-2} \int_{Q_{r/\lambda}} |u_\lambda|^3 \qquad \text{for all } \lambda > 0,$$

where $Q_{r/\lambda}$ is the preimage of Q_r via the mapping $(x, t) \mapsto (\lambda x + x_0, \lambda^2 t + t_0)$.

In what follows we will consider the following scale-invariant quantities:

$$\overline{P}(\rho) := \frac{1}{\rho^2} \int_{Q_\rho} |p - (p)_\rho|^{3/2}, \qquad P(\rho) := \frac{1}{\rho^2} \int_{Q_\rho} |p|^{3/2}, \qquad C(\rho) := \frac{1}{\rho^2} \int_{Q_\rho} |u|^3,$$

(2.3)

and

$$A(\rho) := \frac{1}{\rho} \sup_{t \in (-\rho^2, 0)} \int_{B_\rho} |u(t)|^2, \qquad E(\rho) := \frac{1}{\rho} \int_{Q_\rho} |\nabla u|^2, \qquad (2.4)$$

where we denote by $(f)_\rho$ the average of a function f over a ball B_ρ, that is

$$(f)_\rho := \frac{1}{|B_\rho|} \int_{B_\rho} f. \qquad (2.5)$$

We point out that any L^q norm of an average of any function does not exceed the L^q of the function itself, namely

$$\|(f)_\rho\|_{L^q(B_\rho)} \leq \|f\|_{L^q(B_\rho)}. \qquad (2.6)$$

Thus in particular

$$\overline{P}(\rho) \leq c P(\rho) \qquad \text{for all } \rho > 0.$$

2.1.2 Interpolation Inequality

It turns out that $C(\rho)$ can be estimated in terms of $A(\rho)$ and $E(\rho)$.

Lemma 2.1 (Interpolation inequality) *For all $\rho > 0$*

$$C(\rho) \leq c \left(A(\rho) + E(\rho) \right)^{3/2}. \qquad (2.7)$$

Proof Due to the scale invariance of (2.7), it is sufficient to prove it for $\rho = 1$, in which case we write $Q_1 = Q = I \times B$, where $I = (-1, 0)$, $B = B(0, 1)$ for simplicity. From Lebesgue interpolation and the Sobolev embedding $H^1(B) \subset L^6(B)$ (see Theorem 4.9 in Evans & Gariepy 2015 for a simple proof), we obtain

$$\begin{aligned} \|u\|_{L^3(B)} &\leq \|u\|_{L^6(B)}^{1/2} \|u\|_{L^2(B)}^{1/2} \\ &\leq c \|u\|_{H^1(B)}^{1/2} \|u\|_{L^2(B)}^{1/2} \\ &\leq c \|u\|_{L^2(B)} + c \|\nabla u\|_{L^2(B)}^{1/2} \|u\|_{L^2(B)}^{1/2} \end{aligned}$$

for every time t (which we omitted in the notation), where we used Young's inequality $ab \leq c(a^2 + b^2)$ in the last step. Raising both sides to the power of 3 and integrating over I, we obtain

$$C(1) = \int_Q |u|^3$$

$$\leq c \int_I \left(\int_B |u(t)|^2 \right)^{3/2} dt + c \int_I \left(\int_B |\nabla u(t)|^2 \right)^{3/4} \left(\int_B |u(t)|^2 \right)^{3/4} dt$$

$$\leq c \int_I \left(\sup_{s \in I} \int_B |u(s)|^2 \right)^{3/2} dt + c \left(\sup_{s \in I} \int_B |u(s)|^2 \right)^{3/4} \left(\int_Q |\nabla u|^2 \right)^{3/4}$$

$$= cA(1)^{3/2} + cA(1)^{3/4} E(1)^{3/4},$$

$$\leq cA(1)^{3/2} + cE(1)^{3/2},$$

where we applied Hölder's inequality to the integral $\int_I \left(\int_B |u|^2 \right)^{3/4}$ in the third step. $\qquad\qquad\qquad\qquad\qquad\qquad\qquad\qquad\qquad\qquad\qquad\qquad\qquad\square$

We point out that statement of the lemma has nothing to do with the Navier–Stokes equations (or Navier–Stokes inequality). It is simply an interpolation between different functional spaces.

2.1.3 The Pressure Estimates

We now state two pressure estimates, which bound $\overline{P}$ and P at a scale ρ in terms of quantities involving either u at scales between 2ρ and r or p at scale r. Namely,

$$\overline{P}(\rho) \leq c\, C(2\rho) + c\, \rho^{9/2} \left\{ \sup_{s-\rho^2 < t < s} \int_{2\rho < |x-y| < r} \frac{|u(x,t)|^2}{|x-y|^4} dx \right\}^{3/2} + c \left(\frac{\rho}{r} \right)^{5/2} \left(\overline{P}(r) + C(r) \right) \tag{2.8}$$

and

$$P(\rho)^{4/3} \leq c \left(\frac{\rho}{r} \right)^{-2} A(r)E(r) + c \left(\frac{\rho}{r} \right)^{4/3} P(r)^{4/3} \tag{2.9}$$

for all $0 < \rho < r/2$. (Recall that (y, s) denotes the centre of the upper lid of Q_ρ, see (2.2).) The pressure estimates follow from the partial differential equation

$$-\Delta p = \sum_{i,j=1}^{3} \partial_{ij}(u_i u_j),$$

which is satisfied by p and from the fact that u is divergence-free. In fact, the only tools used in the proof of (2.8) and (2.9) are the properties of the fundamental solution $\Psi(x) := (4\pi |x|)^{-1}$ to the Laplace equation in $\mathbb{R}^3$ as well as some well-known properties of harmonic functions. We give a detailed proof in Sect. 2.5 at the end of this chapter.

We will use (2.8) in the proof of (i) below and (2.9) in the proof of (ii).

2.2 Partial Regularity I

Here we prove part (i) of the Caffarelli–Kohn–Nirenberg theorem, which we restate for the reader's convenience.

Theorem 2.2 *There exists $\varepsilon_0 \in (0, 1)$ such that for any weak solution of the Navier–Stokes inequality on $U \times (a, b)$ (recall Definition 1.6) and any cylinder $Q_r \subset U \times (a, b)$ the condition*

$$\frac{1}{r^2} \int_{Q_r} |u|^3 + |p|^{3/2} \leq \varepsilon_0 \tag{2.10}$$

implies that $u \in L^\infty(Q_{r/2})$.

Proof Due to the scale invariance of (2.10), we can assume (without loss of generality) that $Q_r = Q_r(0, 0)$ and that $r = 1$. Fix $(y, s) \in Q_{1/2}(0, 0)$ and set

$$r_n := 2^{-n}.$$

In the rest of the proof, we will use the shorthand notation

$$Q_r \equiv Q_r(y, s), \qquad B_r \equiv B_r(y).$$

In steps 1–3, we will use induction to show that the claims[1]

$$C(r_n) \leq \varepsilon_0^{2/3} r_n^3, \tag{A_n}$$

$$A(r_n) + E(r_n) \leq c_* \varepsilon_0^{2/3} r_n^2, \tag{B_n}$$

where $c_* > 0$ is an absolute constant (which we fix in (2.15) below), hold for all $n \geq 1$ if ε_0 is sufficiently small. (Recall that $C(\rho) := \rho^{-2} \int_{Q_\rho} |u|^3$, $A(\rho) := \rho^{-1} \sup_{t \in (-\rho^2, 0)} \int_{B_\rho} |u(t)|^2$ and $E(\rho) := \rho^{-1} \int_{Q_\rho} |\nabla u|^2$, see (2.3), (2.4).) In the last step (Step 4), we prove the claim of the theorem using (B_n), $n \geq 1$.

Step 1. We observe that (A_1) holds.
   Indeed, by assumption

$$C(1/2) = 4 \int_{Q_{1/2}} |u|^3 \leq 4\varepsilon_0 \leq \varepsilon_0^{2/3}/8,$$

provided ε_0 is sufficiently small (so that $\varepsilon_0^{1/3} \leq 1/32$).

---

[1] We note that at this point we simplify the classical proof of the theorem, which includes a pressure term in (A_n), see, for example, p. 292 in Robinson et al. (2016), where the left-hand side includes an additional term $r_n^{1/2} \overline{P}(r_n)$.

Step 2. We observe that (B_n) implies (A_{n+1}).

Indeed, (A_{n+1}) follows as in the previous step, but with the assumption replaced by (B_n) and the interpolation inequality (2.7). Namely,

$$C(r_{n+1}) = r_{n+1}^{-2} \int_{Q_{r_{n+1}}} |u|^3 \leq 4 r_n^{-2} \int_{Q_{r_n}} |u|^3 = 4 C(r_n)$$

$$\leq c \, (A(r_n) + E(r_n))^{3/2} \leq c \, c_*^{3/2} \varepsilon_0 r_n^3 \leq \varepsilon_0^{2/3} r_n^3,$$

provided ε is chosen sufficiently small so that

$$\varepsilon_0^{1/3} c \, c_*^{3/2} \leq 1. \tag{2.11}$$

(Recall that $c > 1$ denotes an absolute constant, whose value may change from line to line.)

Step 3. We show that $(A_1), \dots, (A_n), (B_1), \dots, (B_{n-2})$ imply (B_n).

(Note this step, together with steps 1, 2, complete the induction.)

The crucial point of the proof of this step is to use the local energy inequality (2.1) with an appropriately chosen test function. Namely, we let $\phi \in C_0^\infty(Q_{1/3}; [0, \infty))$ be such that $|\partial_t \phi + \Delta \phi| \leq c$ in $Q_{1/3}$ and

$$\frac{1}{c} r_k^{-3} \leq \phi \leq c r_k^{-3}, \quad |\nabla \phi| \leq c r_k^{-4} \quad \text{in} \begin{cases} Q_{r_k} \setminus Q_{r_{k+1}} & \text{for } k = 1, \dots, n-1, \text{ and} \\ Q_{r_k} & \text{for } k = n. \end{cases}$$
$$\tag{2.12}$$

For example, one can take

$$\phi(x, t) := \chi(x, t) \Phi(x, r_n^2 - t),$$

where $\chi \in C_0^\infty(Q_{1/2}; [0, \infty))$ is such that $\chi = 1$ on $Q_{1/4}$ and $\Phi(x, t) = (4\pi t)^{3/2}$ $\exp(-|x|^2/4t)$ denotes the heat kernel. That such ϕ satisfies the above properties can be shown directly by using the fact that $\partial_t \phi + \Delta \phi = 0$ in $Q_{1/4}$ and the rapid decay of the exponential function (particularly that $\alpha^N \exp(\alpha)$ is a bounded function of α for any N). We refer the reader to Lemma 15.11 in Robinson et al. (2016) for a detailed proof of the above properties.

Applying such ϕ in the local energy inequality (2.1) gives

$$\int_{B_{r_n}} |u(t)|^2 + \int_{s-r_n^2}^t \int_{B_{r_n}} |\nabla u|^2 \leq c r_n^3 \int_{s-1/4}^t \int_{B_{1/2}} \left(|u|^2 (\partial_t \phi + \nu \Delta \phi) + (|u|^2 + 2p)(u \cdot \nabla)\phi \right)$$

$$\leq c r_n^3 \left(\int_{Q_{1/2}} |u|^2 + \int_{Q_{1/2}} |u|^3 |\nabla \phi| + \int_{s-1/4}^t \int_{B_{1/2}} p \, u \cdot \nabla \phi \right)$$

$$=: c r_n^3 (I_1 + I_2 + I_3)$$
$$\tag{2.13}$$

for all $t \in (s - r_n^2, s)$. We will show that $(A_1), \dots, (A_n), (B_1), \dots, (B_{n-2})$ give

$$I_1, I_2 \leq c\,\varepsilon_0^{2/3} \qquad \text{and} \qquad I_3 \leq c\,\varepsilon_0^{2/3}(1 + \varepsilon_0^{1/3}c_*^{3/2}), \qquad (2.14)$$

independently of t, where c_* is the constant from (B_n), which we have not fixed yet. By choosing c_* appropriately, we claim that (2.14) finishes this step. Indeed, inserting these estimates into (2.13) and dividing by r_n we arrive at

$$\frac{1}{r_n}\int_{B_{r_n}} |u(t)|^2 + \frac{1}{r_n}\int_{s-r_n^2}^{t}\int_{B_{r_n}} |\nabla u|^2 \leq c\,\varepsilon_0^{2/3}r_n^2(1 + \varepsilon_0^{1/3}c_*^{3/2}).$$

By neglecting the first term on the left-hand side and taking $t \to s$ and then, in turn, neglecting the second term on the left-hand side and taking $\sup_{t\in(s-r_n^2,s)}$, we obtain

$$A(r_n) + E(r_n) \leq c\,\varepsilon_0^{2/3}r_n^2(1 + \varepsilon_0^{1/3}c_*^{3/2}).$$

We now fix

$$c_* := 2c, \qquad (2.15)$$

where c is from the last inequality, and take ε_0 sufficiently small so that (2.11) holds. This gives in particular that $\varepsilon_0^{1/3}c_*^{3/2} \leq 1$, and so $A(r_n) + E(r_n) \leq c_*\varepsilon_0^{2/3}r_n^2$, as required.

It remains to prove (2.14). The estimate on I_1 follows from Hölder's inequality and the assumption (2.10),

$$I_1 \leq |Q_{1/2}|^{1/3}\left(\int_{Q_{1/2}} |u|^3\right)^{2/3} \leq c\varepsilon_0^{2/3}.$$

The estimate on I_2 follows by splitting $Q_{1/2}$ into layers:

$$
\begin{aligned}
I_2 &= \int_{Q_{r_n}} |u|^3|\nabla\phi| + \sum_{k=1}^{n-1}\int_{Q_{r_k}\setminus Q_{r_{k+1}}} |u|^3|\nabla\phi| \\
&\leq cr_n^{-4}\int_{Q_{r_n}} |u|^3 + c\sum_{k=1}^{n-1}r_k^{-4}\int_{Q_{r_k}\setminus Q_{r_{k+1}}} |u|^3 \\
&\leq c\sum_{k=1}^{n}r_k^{-4}\int_{Q_{r_k}} |u|^3 \\
&= c\sum_{k=1}^{n}r_k^{-2}C(r_k) \\
&\leq c\varepsilon_0^{2/3}\sum_{k=1}^{n}r_k \\
&\leq c\varepsilon_0^{2/3},
\end{aligned}
$$

where we used the bound on $|\nabla\phi|$ from (2.12) as well as $(A_1), \ldots, (A_n)$.

As for I_3, due to the presence of the pressure p, we will use a slightly different splitting. The point is that we want to use the pressure estimate (2.8), which allows us to estimate only $p - (p)_r$ on a cylinder Q_r. In order to incorporate the average $(p)_r$ into I_3, we will use the fact that

$$\fint_{B_{1/2}} f(t)u(t) \cdot \nabla\phi(t) = 0 \tag{2.16}$$

for all $t \in (s - 1/4, s)$ and any f which is a function of t only. Identity (2.16) follows by integration by parts and using the facts that u is divergence-free and that $\phi(t)$ has compact support in $B_{1/3}$ for every $t \in (-2^{-4}, 0]$.

To be more precise we let $\chi_k \in C_0^\infty(Q_{r_k}; [0, 1])$ be such that $\chi_k = 1$ on $Q_{7r_k/8}$ and $|\nabla\chi_k| \le c/r_k$. Observe that

$$|\nabla((\chi_k - \chi_{k+1})\phi)| \le c r_k^{-4} \quad \text{in } Q_{r_k} \setminus Q_{r_{k+1}}, k = 1, \ldots, n-1, \tag{2.17}$$

and

$$|\nabla(\chi_n\phi)| \le c r_n^{-4} \quad \text{in } Q_{r_n}, \tag{2.18}$$

by the chain rule and the estimates on $\phi, |\nabla\phi|$ from (2.12). Since

$$\operatorname{supp}\phi \subset Q_{1/3} \subset Q_{7/16} \subset \{\chi_1 = 1\},$$

we see that $\chi_1\phi = \phi$, and so ϕ can be decomposed as

$$\phi = \chi_n\phi + \sum_{k=1}^{n-1}(\chi_k - \chi_{k+1})\phi.$$

Plugging this in the definition of I_3, we obtain for all $t \in (s - r_n^2, s)$

$$
\begin{aligned}
I_3 &= \int_{s-1/4}^t \int_{B_{1/2}} p\, u \cdot \nabla(\chi_n\phi) + \sum_{k=1}^{n-1} \int_{s-1/4}^t \int_{B_{1/2}} p\, u \cdot \nabla((\chi_k - \chi_{k+1})\phi) \\
&= \int_{s-1/4}^t \int_{B_{1/2}} (p - (p)_{r_n})u \cdot \nabla(\chi_n\phi) + \sum_{k=1}^{n-1} \int_{s-1/4}^t \int_{B_{1/2}} (p - (p)_{r_k})u \cdot \nabla((\chi_k - \chi_{k+1})\phi),
\end{aligned}
$$

where we used (2.16) in the second line. We can now use the fact that $\operatorname{supp}\chi_k \subset Q_{r_k}$ and the estimates (2.17), (2.18) to obtain

$$I_3 \leq c r_n^{-4} \int_{Q_{r_n}} |(p - (p)_{r_n})u| + c \sum_{k=1}^{n-1} r_k^{-4} \int_{Q_{r_k}} |(p - (p)_{r_k})u|$$

$$= c \sum_{k=1}^{n} r_k^{-4} \int_{Q_{r_k}} |(p - (p)_{r_k})u|$$

$$\leq c \sum_{k=1}^{n} r_k^{-4} \left(\int_{Q_{r_k}} |u|^3 \right)^{1/3} \left(\int_{Q_{r_k}} |p - (p)_{r_k}|^{3/2} \right)^{2/3} \qquad (2.19)$$

$$= c \sum_{k=1}^{n} r_k^{-2} C(r_k)^{1/3} \overline{P}(r_k)^{2/3}$$

$$\leq c \varepsilon_0^{2/9} \sum_{k=1}^{n} r_k^{-1} \overline{P}(r_k)^{2/3},$$

where we used Hölder's inequality in the third line and (A_k) to bound $C(r_k)$ ($k = 1, \ldots, n$) in the last line.

Applying (2.8) with $r = 1/2$ and $\rho = r_k$ (for $k \geq 2$), we obtain

$$\overline{P}(r_k) \leq c C(r_{k-1}) + c r_k^{9/2} \left\{ \sup_{s - r_k^2 < t < s} \int_{r_{k-1} < |x-y| < 1/2} \frac{|u(x,t)|^2}{|x-y|^4} dx \right\}^{3/2} + c r_k^{5/2} \left(\overline{P}(1/2) + C(1/2) \right)$$

$$\leq c \varepsilon_0^{2/3} r_k^3 + c r_k^{9/2} \left\{ \sum_{l=1}^{k-2} \sup_{s - r_l^2 < t < s} \int_{r_{l+1} < |x-y| < r_l} \frac{|u(x,t)|^2}{|x-y|^4} dx \right\}^{3/2} + c r_k^{5/2} \varepsilon_0$$

$$\leq c \varepsilon_0^{2/3} r_k^{5/2} + c r_k^{9/2} \left\{ \sum_{l=1}^{k-2} r_l^{-4} \sup_{s - r_l^2 < t < s} \int_{r_{l+1} < |x-y| < r_l} |u(t)|^2 \right\}^{3/2}$$

$$\leq c \varepsilon_0^{2/3} r_k^{5/2} + c r_k^{9/2} \left\{ \sum_{l=1}^{k-2} r_l^{-3} A(r_l) \right\}^{3/2}$$

$$\leq c \varepsilon_0^{2/3} r_k^{5/2} + c r_k^{9/2} \left\{ c_* \varepsilon_0^{2/3} \sum_{l=1}^{k-2} r_l^{-1} \right\}^{3/2}$$

$$= c \varepsilon_0^{2/3} r_k^{5/2} + c \varepsilon_0 c_*^{3/2} r_k^{9/2} \left\{ 2^{k-1} - 2 \right\}^{3/2}$$

$$\leq c \varepsilon_0^{2/3} r_k^{5/2} (1 + \varepsilon_0^{1/3} c_*^{3/2}),$$

where we also used the fact that $r_k = 2^{-k}$ (so that in particular the supremum in the second line is taken over a larger interval than in the first line; and also $r_k \leq 1$, which we used in the third and last lines), as well as $(B_1), \ldots, (B_{k-2})$. As for the case $k = 1$ (for which (2.8) does not apply), we obtain the same estimate directly from the assumption (2.10),

$$\overline{P}(1) = 4 \int_{Q_{1/2}} |p - (p)_{1/2}|^{3/2} \leq c \int_{Q_{1/2}} |p|^{3/2} \leq c \varepsilon_0 < c \varepsilon_0^{2/3} r_1^{5/2} (1 + \varepsilon_0^{1/3} c_*^{3/2}).$$

Plugging these into (2.19), we obtain

$$I_3 \le c\varepsilon_0^{2/3}(1 + \varepsilon_0^{1/3}c_*^{3/2})^{2/3} \sum_{k=1}^{n} r_k^{2/3} \le c\varepsilon_0^{2/3}(1 + \varepsilon_0^{1/3}c_*^{3/2}),$$

as required.

Step 4. We use (B_n), $n \ge 1$, to prove the claim of the theorem.
From (B_n),

$$\sup_{t \in (s-r_n^2, s)} \frac{1}{|B_{r_n}|} \int_{B_{r_n}(y)} |u(t)|^2 = \frac{c}{r_n^2} A(r_n) \le cc_* \varepsilon_0^{2/3} = c \qquad \text{for } (y, s) \in Q_{1/2}(0, 0), n \ge 1.$$

Thus (by varying $s \in (-1/4, 0)$)

$$\frac{1}{|B_{r_n}|} \int_{B_{r_n}(y)} |u(t)|^2 \le c \qquad \text{for } t \in (-1/4, 0), \ y \in B_{1/2}(0), n \ge 1.$$

Using Lebesgue Differentiation Theorem (at the limit $n \to \infty$), we obtain

$$|u(y, t)|^2 \le c \qquad \text{for } t \in (-1/4, 0) \text{ and a.e. } y \in B_{1/2}(0),$$

that is $u \in L^\infty(Q_{1/2}(0, 0))$, as required.   □

2.3 Partial Regularity II

Here, we prove part (ii) of the Caffarelli–Kohn–Nirenberg theorem, which we restate for reader's convenience.

Theorem 2.3 *There exists $\varepsilon_1 > 0$ such that whenever (u, p) is a weak solution of the Navier–Stokes inequality (on an open set $U \times (a, b)$), $(y, s) \in U \times (a, b)$ and*

$$\limsup_{r \to 0} \frac{1}{r} \int_{Q_r} |\nabla u|^2 \le \varepsilon_1 \qquad (2.20)$$

(here $Q_r = Q_r(y, s)$) then $u \in L^\infty(Q_\rho)$ for some $\rho > 0$.

Before the proof, we note that we will use another scale-invariant quantity (in addition to $A(r)$, $E(r)$, $P(r)$, $C(r)$, recall (2.3), (2.4)), namely,

$$G(r) := \frac{1}{r^2} \int_{Q_r} \left||u|^2 - \left(|u|^2\right)_r\right|^{3/2}.$$

Recall that $(\cdot)_r$ denotes average over B_r (see (2.5)). As in Lemma 2.1, we will need an interpolation inequality for $G(r)$.

Lemma 2.4 *For every $\rho > 0$,*

$$G(\rho) \le cA(\rho)^{3/4}E(\rho)^{3/4}. \tag{2.21}$$

Observe that, in contrast to Lemma 2.1, we obtain a product of A and E on the right-hand side (rather than a sum). This will be crucial in the proof of the theorem (which we also point out in the informal discussion following (2.28)).

Proof For almost every $t \in (s - \rho^2, 2)$

$$
\begin{aligned}
\int_{B_\rho} \left| |u|^2 - \left(|u|^2 \right)_\rho \right|^{3/2} &\le c \left(\int_{B_\rho} |\nabla |u|^2| \right)^{3/2} \\
&\le c \left(\int_{B_\rho} |u|\,|\nabla u| \right)^{3/2} \\
&\le c \|u\|^{3/2} \|\nabla u\|^{3/2},
\end{aligned}
$$

where we used the Poincaré inequality $\|f - (f)_\rho\|_{L^{3/2}(B_\rho)} \le c\|\nabla f\|_{L^1(B_\rho)}$ (see Theorem 4.9 in Evans and Gariepy (2015) for a proof) in the first step and the Cauchy–Schwarz inequality in the last step. Integrating in $\int_{s-\rho^2}^s dt$, dividing by ρ^2, and applying Hölder's inequality (with exponents 4, 4/3) give (2.21). $\qquad\square$

Moreover, we will also use a similar trick as in (2.16) above, namely, we can rewrite the local energy inequality (2.1) in the form

$$\int_U |u(t)|^2 \phi(t) + 2 \int_a^t \int_U |\nabla u|^2 \phi \le \int_a^t \int_U \left(|u|^2 (\partial_t \phi + \Delta\phi) + (|u|^2 + 2p + f)(u \cdot \nabla)\phi \right), \tag{2.22}$$

where $\phi \in C_0^\infty(U \times (a, b); [0, \infty))$, $t \in (a, b)$ and f is any function of time.

We can now prove the theorem.

Proof of Theorem 2.3 Without loss of generality, we can assume that $(y, s) = (0, 0)$ and that for sufficiently small r, say $r \in (0, 1)$, $Q_r = Q_r(0, 0) \subset U \times (a, b)$. We will show that

$$C(\rho) + P(\rho) < \varepsilon_0 \tag{2.23}$$

for some $\rho > 0$, where ε_0 is from the first partial regularity theorem (Theorem 2.2), which then gives $u \in L^\infty(Q_{\rho/2})$, proving the claim.

Step 1. We show that

$$A(\theta r) + E(\theta r) \le c\theta^2 (A(r) + E(r)) + c\theta^{-6} \left(A(r)E(r) + P(r)^{4/3} \right) \tag{2.24}$$

for all $r, \theta \in (0, 1)$.

To this end we, will use the local energy inequality (2.22) with a carefully chosen test function ϕ. Namely, let $\phi \in C_0^\infty(Q_r; [0, \infty))$ be such that

$$\phi \geq c(\theta r)^{-1} \quad \text{in } Q_{\theta r} \tag{2.25}$$

and

$$\phi \leq c(\theta r)^{-1}, \ |\nabla \phi| \leq c(\theta r)^{-2}, \ |\partial_t \phi + \Delta \phi| \leq c\theta^2 r^{-3} \quad \text{in } Q_r. \tag{2.26}$$

One can take for example

$$\phi(x, t) := (\theta r)^2 \chi(x, t) \Phi(x, (\theta r)^2 - t),$$

where $\chi \in C_0^\infty(Q_r; [0, 1])$ is such that $\chi = 1$ on $Q_{r/2}$ and as before $\Phi(x, t) = (4\pi t)^{3/2} \exp(-|x|^2/4t)$ denotes the heat kernel. As before, properties (2.25), (2.26) follow by a direct calculation (see, for example, Lemma 16.6 in Robinson et al. 2016).

Applying such ϕ in the local energy inequality (2.22) with $f := -\left(|u|^2\right)_r$ gives

$$\frac{1}{\theta r} \int_{B_{\theta r}} |u(t)|^2 + \frac{2}{\theta r} \int_{-(\theta r)^2}^t \int_{B_{\theta r}} |\nabla u|^2 \leq c\theta^2 r^{-3} \int_{-r^2}^t \int_{B_r} |u|^2$$
$$+ c(\theta r)^{-2} \int_{-r^2}^t \int_{B_r} \left| u \left(|u|^2 - \left(|u|^2 \right)_r + 2p \right) \right|$$

for $t \in (-(\theta r)^2, 0)$, where we applied the lower bound (2.25) on the left-hand side and (2.26) on the right-hand side. As before, we first neglect the first term on the left-hand side and take $t \to 0$, and then neglect the second term and take $\sup_{t \in (-(\theta r)^2, 0)}$, to obtain

$$A(\theta r) + E(\theta r) \leq c\theta^2 r^{-3} \int_{Q_r} |u|^2 + c(\theta r)^{-2} \int_{Q_r} \left| u \left(|u|^2 - \left(|u|^2 \right)_r + 2p \right) \right|$$

$$\leq c\theta^2 r^{-4/3} \left(\int_{Q_r} |u|^3 \right)^{2/3}$$

$$+ c\theta^{-2} r^{-2} \left(\int_{Q_r} |u|^3 \right)^{1/3} \left[\left(\int_{Q_r} \left| |u|^2 - \left(|u|^2 \right)_r \right|^{3/2} \right)^{2/3} + \left(\int_{Q_r} |p|^{3/2} \right)^{2/3} \right]$$

$$= c\theta^2 C(r)^{2/3} + c\theta^{-2} C(r)^{1/3} \left[G(r)^{2/3} + P(r)^{2/3} \right]$$

$$\leq c\theta^2 C(r)^{2/3} + c\theta^{-6} \left[G(r)^{4/3} + P(r)^{4/3} \right]$$

$$\leq c\theta^2 \left(A(r) + E(r) \right) + c\theta^{-6} \left(A(r)E(r) + P(r)^{4/3} \right)$$

where we used (2.16) in the second line, Hölder's inequality (twice, both with exponents $3, 3/2$) in the third line, Young's inequality $ab \leq c\theta^4 a^2 + c\theta^{-4} b^2$ in the penultimate line and the interpolation inequalities (2.7) and (2.21) in the last line.

Step 2. We show that

$$H(\theta r) \leq cE(r) + cH(r)\left(\theta^{-9}E(r) + \theta\right) \tag{2.27}$$

for all $r, \theta \in (0, 1/2)$, where

$$H(r) := A(r) + E(r) + \theta^{-7}P(r)^{4/3}. \tag{2.28}$$

Before showing (2.27) we first comment on the idea of considering such H in an informal way.

In view of the assumption (2.20) (which implies that $E(r) \leq \varepsilon_1$ for sufficiently small r) and (2.24), we are interested in choosing $\theta, \varepsilon_1 \in (0, 1/2)$ such that the right-hand side of (2.24) is small, so that the quantities A and E at a smaller radius θr are much smaller than at radius r. The first term on the right-hand side of (2.24) can be made small by choosing θ small. As for the second term, its first ingredient (i.e. $\theta^{-6}A(r)E(r)$) can be made small by choosing ε_1 sufficiently small to compensate the size of θ^{-6}. Its second ingredient, $\theta^{-6}P(r)^{4/3}$, however, does not seem easy to make small. In fact, at this point its appearance might seem like a fatal blow to our argument.

A rescue comes from the pressure estimate (2.9). Namely, by taking $\rho = \theta r$ in (2.9), we obtain

$$P(\theta r)^{4/3} \leq c\theta^{-2}A(r)E(r) + c\theta^{4/3}P(r)^{4/3}. \tag{2.29}$$

The essential property of this estimate is that negative power of θ appears only at the term $A(r)E(r)$, which (as above) we can make small by taking ε_1 small. In order to see how this property deals with the term $\theta^{-6}P(r)^{4/3}$, one can simply consider

$$H(r) := A(r) + E(r) + \theta^\alpha P(r)^{4/3},$$

where α is any number smaller than -6. Then

$$\theta^{-6}P(r)^{4/3} \leq \theta^{-6-\alpha}H(r),$$

and since $-6 - \alpha > 0$ the last term can be made small by taking θ small. For simplicity, we will take

$$\alpha := -7,$$

and thus we arrive at (2.28). The above strategy of "making the terms of the right-hand side small" is made rigorous by exploiting (2.27) in the steps below.

We now briefly show (2.27). For simplicity, we omit the argument "r", that is, we write

$$A = A(r), \quad E = E(r), \quad P = P(r), \quad H = H(r).$$

Adding (2.24) to (2.29) multiplied by θ^{-7}, we obtain

$$H(\theta r) \leq c\,\theta^2(A+E) + c\,\theta^{-6}(AE + P^{4/3}) + c\,\theta^{-9}AE + c\,\theta^{-17/3}P^{4/3}$$
$$\leq c\,\theta^2(H+E) + c\,\theta^{-6}(HE + \theta^7 H) + c\,\theta^{-9}HE + c\,\theta^{4/3}H$$
$$= c\,\theta^2 E + c\,H\left(E\left(\theta^{-6}+\theta^{-9}\right)+\theta^2+\theta+\theta^{4/3}\right)$$
$$\leq c\,E + c\,H\left(E\theta^{-9}+\theta\right),$$

where we used the definition of H (particularly the facts $A \leq H$ and $P^{4/3} \leq \theta^7 H$) in the second line, and the fact that $\theta < 1$ in the last line.

Step 3. We show that for some choice of θ, ε_1, $R \in (0,1)$ the assumption (2.20) gives

$$H(\theta r) \leq \frac{\varepsilon_0}{8c_\star} + \frac{1}{2}H(r) \qquad \text{for all } r \in (0,R), \tag{2.30}$$

where $c_\star > 1$ is the constant from the interpolation inequality (2.7) (its appearance here will be clear in the next step).

The assumption (2.20) yields the existence of $R \in (0,1)$ such that $E(r) \leq \varepsilon_1$ for $r \in (0,R)$. Step 2 gives

$$H(\theta r) \leq c\left[\varepsilon_1 + H(r)\left(\varepsilon_1\theta^{-9}+\theta\right)\right] \quad \text{for } r \in (0,R). \tag{2.31}$$

We now fix $\theta \in (0,1)$ such that

$$\theta \leq \frac{1}{4c} \tag{2.32}$$

and then we fix $\varepsilon_1 \in (0,1)$ such that

$$\varepsilon_1 \leq \frac{\varepsilon_0}{8c_\star c}, \qquad \varepsilon_1 \leq \frac{\theta^9}{4c},$$

where c is from (2.31), which gives (2.30).

Step 4. We deduce (2.23) from Step 3.

We iterate (2.30) with $r = \theta^k R$, $k \geq 0$, to obtain

$$H(\theta^k R) \leq \frac{\varepsilon_0}{8c_\star}\sum_{l=0}^{k-1}2^{-l} + 2^{-k}H(R) \leq \frac{\varepsilon_0}{4c_\star} + 2^{-k}H(R) \qquad \text{for } k \geq 0,$$

and so

$$H(\theta^k R) \leq \frac{\varepsilon_0}{2c_\star}$$

for sufficiently large k. Thus, by applying the interpolation inequality (2.7) and using the facts that ε_0, $\theta < 1$ and $c_\star > 1$ we obtain for such k

$$C(\theta^k R) + P(\theta^k R) \leq c_\star \left(A(\theta^k R) + E(\theta^k R)\right)^{3/2} + P(\theta^k R)$$

$$\leq c_\star H(\theta^k R)^{3/2} + \theta^7 H(\theta^k R)$$

$$\leq c_\star \left(\frac{\varepsilon_0}{2c_\star}\right)^{3/2} + \theta^7 \frac{\varepsilon_0}{2c_\star}$$

$$< \frac{\varepsilon_0}{2} + \frac{\varepsilon_0}{2}$$

$$= \varepsilon_0,$$

which is (2.23) with $\rho = \theta^k R$. □

Corollary 2.5 *If (u, p) is a weak solution of the Navier–Stokes inequality and*

$$\limsup_{r \to 0} \sup_{t \in (s - r^2, s)} \frac{1}{r} \int_{B_r(y)} |u(t)|^2 \leq \varepsilon_1$$

then $u \in L^\infty(Q_\rho(y, s))$ for some $\rho > 0$.

Proof The claim follows in the same way as Theorem 2.3 after switching "A" with "E" in steps 2–4. □

2.4 Bounds on the Dimension of the Singular Set

Here, we prove Corollary 1.5, that is, given a weak solution (u, p) to the Navier–Stokes inequality on an open set $U \times (a, b)$ we let

$$S := \{(x, t) \in U \times (a, b) \colon u \text{ is unbounded in any neighbourhood of } (x, t)\},$$

and we show that for every compact set $K \subset U \times (a, b)$

$$d_B(S \cap K) \leq 5/3 \tag{2.33}$$

and that

$$d_H(S) \leq 1. \tag{2.34}$$

In fact, we will show a stronger property than (2.34),

$$\mathcal{P}^1(S) = 0, \tag{2.35}$$

where $\mathcal{P}^1$ denotes the one-dimensional parabolic Hausdorff measure, that is (1.14) with balls B_k replaced by cylinders Q_k (note that the partial regularity Theorem 1.4 is concerned with cylinders, rather than (four-dimensional) balls. In order to see that (2.35) implies $d_H(S) \leq 1$ note that $Q_k \subset 2B_k$ (if the radius is ≤ 1). Thus $\mathcal{P}^1(S) = 0$ implies $\mathcal{H}^1(S) = 0$, which in particular gives (2.34).

As compared to the non-anticipating cylinders $Q_r(y, s) = B_r(y) \times (s - r^2, s]$ (recall (2.2)), which we have considered so far, in this section we will denote a cylinder *centred* at (y, s) by

$$Q_r^* = Q_r^*(y, s) = B_r(y) \times (s - r^2, s + r^2).$$

We consider such cylinders since both the definitions of d_B (recall (1.12)) and $\mathcal{P}^1$ are concerned with covers of a set. In this context, it is more reasonable to consider the notion of a cylinder $Q_r^*(y, s)$ that contains (y, s) in its interior.

We first prove (2.33).

Lemma 2.6 *Let (u, p) be a weak solution of the Navier–Stokes inequality on an open set $U \times (a, b)$ with $p \in L_{loc}^{5/3}(U \times (a, b))$. Then $d_B(S \cap K) \leq 5/3$ for every compact set $K \subset U \times (a, b)$.*

Recall that

$$d_B(S \cap K) := \limsup_{r \to 0} \frac{M(S \cap K, r)}{-\log r}$$

(see (1.12)), where $M(S \cap K, r)$ stands for the maximal number of r-balls with centres in $S \cap K$.

Proof Let $r_0 \in (0, 1)$ be such that $\text{dist}(K, \partial(U \times (a, b))) > r_0$ and note that for every $z = (y, s) \in S \cap K$ and $r \in (0, r_0)$ we have

$$\int_{Q_r^*(z)} |u|^3 + |p|^{3/2} \geq \varepsilon_0 r^2. \tag{2.36}$$

Indeed, if this was not the case then

$$\int_{Q_r(y, s + r^2/8)} |u|^3 + |p|^{3/2} \leq \int_{Q_r^*(y, s)} |u|^3 + |p|^{3/2} < \varepsilon_0 r^2$$

(as $Q_r(y, s + r^2/8) \subset Q_r^*(y, s)$) and Theorem 2.2 would give that $u \in L^\infty(Q_{r/2}(y, s + r^2/8))$, which contradicts the assumption that $z \in S$.

We will show that for sufficiently small r

$$M'(S \cap K, r) \leq c_K r^{5/3}, \tag{2.37}$$

where $M'(S \cap K, r)$ denotes the maximal number of pairwise disjoint r-cylinders Q^* with centres in $S \cap K$. Since $Q_r^* \subset B_{2r}$ for any $r \in (0, 1)$ and any pair of a cylinder Q_r^* and a (four-dimensional) ball B_{2r} with coinciding centres, we have

$$M(S \cap K, 2r) \leq M'(S \cap K, r).$$

Therefore, given (2.37), we obtain the required estimate $d_B(S \cap K) \leq 5/3$ (see the definition of d_B above).

In order to see (2.37), let $Q_1^*, \ldots, Q_{M'(S \cap K,r)}^*$ be pairwise disjoint r-cylinders with centres in $S \cap K$, where $r \in (0, r_0)$. Note that such cylinders are contained in $K + \overline{Q_{r_0}}(0, 0)$, which is a compact subset of $U \times (a, b)$. Thus, using (1.6) and the assumption $p \in L_{loc}^{5/3}(U \times (a, b))$

$$
\begin{aligned}
c_K &\geq \int_{K + \overline{Q_{r_0}}(0,0)} |u|^{10/3} + |p|^{5/3} \\
&\geq \sum_{i=1}^{M'(S \cap K,r)} \int_{Q_i^*} |u|^{10/3} + |p|^{5/3} \\
&\geq c\, r^{-5/9} \sum_{i=1}^{M'(S \cap K,r)} \left(\int_{Q_i^*} |u|^3 + |p|^{3/2} \right)^{10/9} \\
&\geq c\, r^{5/3} \sum_{i=1}^{M'(S \cap K,r)} \varepsilon_0^{10/9} \\
&= c\, \varepsilon_0^{10/9} r^{5/3} M'(S \cap K, r),
\end{aligned}
$$

as required, where we also used Hölder's inequality (with exponents 10/9, 10) and (2.36). $\qquad\square$

We have included the assumption that $p \in L_{loc}^{5/3}$ in the statement of the lemma since it is a well-known regularity of the pressure function corresponding to any Leray–Hopf weak solution (see Corollary 5.2 in Robinson et al. (2016) for the case of the torus $\mathbb{T}^3$ and the whole space $\mathbb{R}^3$, and Theorem 5.7 therein, which is based on the argument due to Sohr and von Wahl (1986), for the case of a smooth bounded domain). We have already pointed out in (1.7) that this is particularly clear in the case of the whole space $U = \mathbb{R}^3$.

However, if $p \notin L_{loc}^{5/3}$ (i.e. if one considers merely suitable weak solutions, for which $p \in L_{loc}^{3/2}$, recall Definition 1.3) then, as in the proof of the lemma above, one obtains the bound $d_B(S \cap K) \leq 2$ (instead of 5/3). In either case $\mathcal{H}^s(S \cap K) = 0$ for all compact K and $s > 2$ (by (1.15)), and so in particular

$$
|S| = 0.
$$

We now turn to a proof of the estimate $\mathcal{P}^1(S) = 0$. To this end, we will need the following simple corollary of Theorem 2.3, where the non-anticipating cylinders Q_r are replaced by the centred cylinders Q_r^*.

Corollary 2.7 *If (u, p) is a weak solution of the Navier–Stokes inequality, then*

$$
\limsup_{r \to 0} \frac{1}{r} \int_{Q_r^*(y,s)} |\nabla u|^2 \leq \varepsilon_1 \tag{2.38}
$$

implies that $u \in L^\infty(Q_\rho^*(y, s))$ for some $\rho > 0$.

Proof The assumption of the corollary gives in particular the assumption (2.20) of Theorem 2.3. It remains to observe that in the proof of Theorem 2.3 we have actually shown that $C(\rho) + P(\rho) < \varepsilon_0$ (for some $\rho > 0$), that is

$$\int_{Q_\rho(y,s)} |u|^3 + |p|^{3/2} < \varepsilon_0 r^2.$$

Therefore, the same is true for the $Q_\rho(y, s + \varepsilon)$ for sufficiently small $\varepsilon > 0$. Theorem 2.2 thus gives that $u \in L^\infty Q_{\rho/2}(y, s + \varepsilon)$, and the claim follows.   □

We will also need the Vitali Covering Lemma in the following form: given a family of cylinders $Q_r^*(x, t)$, there exists a countable (or finite) subfamily $\{Q_{r_i}(x_i, t_i)\}$ of pairwise disjoint cylinders such that for any cylinder $Q_r(x, t)$ in the original family there exists an i such that $Q_r(x, t) \subset Q_{5r_i}(x_i, t_i)$. (For a proof see Lemma 6.1 in Caffarelli et al. 1982.)

We can now prove the result.

Lemma 2.8 *Let (u, p) be a weak solution of the Navier–Stokes inequality on an open set $U \times (a, b)$. Then $\mathcal{P}^1(S) = 0$.*

Proof Note that it is sufficient to find a cover S by cylinders $Q_{r_i}^*(x_i, t_i)$ such that $\sum_i r_i \leq \delta$ for any given $\delta > 0$.

Fix $\delta > 0$ and let V be an open set containing S such that

$$\frac{5}{\varepsilon_1} \int_V |\nabla u|^2 \leq \delta.$$

Such V exists since $|S| = 0$ and $\nabla u \in L^2(U \times (a, b))$ (by the regularity properties of any weak solution of the NSI, recall the beginning of this chapter). For each $(x, t) \in S$, choose $r \in (0, \delta)$ such that $Q_{r/5}^*(x, t) \subset V$ and

$$\frac{5}{r} \int_{Q_{r/5}^*(x,t)} |\nabla u|^2 \geq \varepsilon_1.$$

Such a choice is possible, for otherwise the point (x, t) would be regular due to Corollary 2.7 above. We now use the Vitali Covering Lemma to extract a countable (or finite) disjoint subcollection of these cylinders $\{Q_{r_i/5}(x_i, t_i)\}$ such that the singular set S is still covered by $\{Q_{r_i}(x_i, t_i)\}$. Then

$$\sum_i r_i \leq \frac{5}{\varepsilon_1} \sum_i \int_{Q_{r_i/5}(x,t)} |\nabla u|^2 \leq \frac{5}{\varepsilon_1} \int_V |\nabla u|^2 \leq \delta,$$

as required.   □

2.5 The Pressure Estimates

In this section, we prove the pressure estimates (2.8), (2.9). For this purpose, we recall the well-known Calderón-Zygmund inequality,

$$\|\partial_{ij}\Psi * f\|_{L^q(\mathbb{R}^3)} \le C_q \|f\|_{L^q(\mathbb{R}^3)}, \qquad q \in (1, \infty) \tag{2.39}$$

for all $i, j = 1, 2, 3$, where $\Psi(x) = (4\pi|x|)^{-1}$ stands for the fundamental solution of the Laplace equation in $\mathbb{R}^3$. We refer the reader to Appendix B.2 in Robinson et al. (2016) for a proof of (2.39).

We can now prove the pressure estimate. We start with (2.8). For brevity, we will write $Q_r \equiv Q_r(y, s)$ and $\sum_{i,j} \equiv \sum_{i,j=1}^3$.

Lemma 2.9 (The pressure estimate I) *If $p \in L^{3/2}(Q_r)$ satisfies*

$$-\Delta p = \sum_{i,j} \partial_{ij}(u_i u_j)$$

in Q_r and $0 < \rho \le r/2$ then

$$\overline{P}(\rho) \le c\, C(2\rho) + c\, \rho^{9/2} \left\{ \sup_{s-\rho^2 < t < s} \int_{2\rho < |x-y| < r} \frac{|u(x,t)|^2}{|x-y|^4} dx \right\}^{3/2} + c \left(\frac{\rho}{r}\right)^{5/2} \left(\overline{P}(r) + C(r)\right).$$

Proof Without loss of generality, we will take $(y, s) = (0, 0)$ (so in particular $B_r = B_r(0)$).

Step 1. We reduce the claim to an inequality in space only.

Namely, in the remaining steps, we will show that

$$\|p - (p)_\rho\|_{L^{3/2}(B_\rho)} \le c\|u\|_{L^3(B_{2\rho})}^2 + c\,\rho^3 \int_{2\rho < |x| < r} \frac{|u(x,t)|^2}{|x|^4} dx$$
$$+ c\left(\frac{\rho}{r}\right)^3 \left(\|p - (p)_r\|_{L^{3/2}(B_r)} + \|u\|_{L^3(B_r)}^2\right) \tag{2.40}$$

for each time $t \in (-\rho^2, 0)$. (Recall that $(p)_r$ denotes the average over B_r (rather than over a space–time cylinder).) Then, taking the 3/2 power of (2.40), taking supremum in $t \in (-\rho^2, 0)$ of the second term on the right-hand side, integrating in $\int_{-\rho^2}^0 dt$, and extending the domain of integration on the right-hand side we obtain

$$\int_{Q_\rho} |p - (p)_\rho|^{3/2} \le c \int_{Q_{2\rho}} |u|^3 + c\,\rho^{13/2} \sup_{t\in(-\rho^2,0)} \left\{\int_{2\rho<|x|<r} \frac{|u(x,t)|^2}{|x|^4} dx\right\}^{3/2}$$
$$+ c\left(\frac{\rho}{r}\right)^{9/2} \left(\int_{Q_r} |p - (p)_r|^{3/2} + \int_{Q_r} |u|^3\right),$$

which is the claim of the lemma multiplied by ρ^2.

Step 2. We decompose the pressure function p.

Namely, let $\phi \in C_0^\infty(B_r; [0, 1])$ be such that $\phi \equiv 1$ on $B_{7r/8}$ and

$$p_1 := \sum_{i,j} \partial_{ij}\Psi * (u_i u_j \phi). \tag{2.41}$$

(Recall $\Psi(x) = (4\pi|x|)^{-1}$ stands for the fundamental solution of the Laplace equation in $\mathbb{R}^3$.) Thus letting $p_2 := p - p_1$ we obtain a decomposition of p,

$$p = p_1 + p_2.$$

The point of such a decomposition is that $p_1 - (p_1)_\rho$ can be estimated in the $L^{3/2}$ norm on B_ρ by a simple split of the domain of integration and the use of the Calderón–Zygmund inequality (2.39). This will give us the first two terms appearing on the right-hand side of (2.40), and is presented in Step 3.

Moreover, such a decomposition gives that

$$-\Delta p_1 = \sum_{i,j} \partial_{ij}(u_i u_j \phi) \qquad \text{in } \mathbb{R}^3,$$

and so in particular p_2 is harmonic in $B_{7r/8}$ (as $\phi = 1$ there). In Step 4, we will use interior estimates of harmonic functions to estimate the $L^{3/2}$ norm of $p_2 - (p_2)_\rho$ on B_ρ by the last term on the right-hand side of (2.40).

This will conclude the proof of (2.40) by a simple use of the triangle inequality,

$$\|p - (p)_\rho\|_{L^{3/2}(B_\rho)} \le \|p_1 - (p_1)_\rho\|_{L^{3/2}(B_\rho)} + \|p_2 - (p_2)_\rho\|_{L^{3/2}(B_\rho)}.$$

Step 3. We estimate p_1.

Namely, in this step we will show that

$$\|p_1 - (p_1)_\rho\|_{L^{3/2}(B_\rho)} \le c\|u\|_{L^3(B_{2\rho})}^2 + c\rho^3 \int_{2\rho < |x| < r} \frac{|u(x,t)|^2}{|x|^4} \mathrm{d}x. \tag{2.42}$$

First decompose p_1 into

$$p_1(x) = p_{1,1}(x) + p_{1,2}(x) = \sum_{i,j} \partial_{ij}\Psi * (u_i u_j \phi \chi_{B_{2\rho}}) + \int_{\mathbb{R}^3 \setminus B_{2\rho}} \partial_{ij}\Psi(x-z)u_i(z)u_j(z)\phi(z)\mathrm{d}z.$$

The function $p_{1,1}$ can be bounded using the Calderón–Zygmund inequality (2.39),

$$\|p_{1,1}\|_{L^{3/2}(B_\rho)} \le \|p_{1,1}\|_{L^{3/2}(\mathbb{R}^3)} \le c \left\| \sum_{i,j} u_i u_j \phi \chi_{B_{2\rho}} \right\|_{L^{3/2}(\mathbb{R}^3)} \le c \|u\|^2_{L^3(B_{2\rho})}.$$

As for $p_{1,2}$ observe that for any $k \ge 0$

$$|D^k \Psi(x)| \le c|x|^{-(1+k)} \tag{2.43}$$

Thus

$$|p_{1,2}(x)| \le c \int_{2\rho < |z| < r} \frac{|u(z)|^2 \phi(z)}{|x-z|^3} dz \le c \int_{2\rho < |z| < r} \frac{|u(z)|^2}{|z|^3} dz \qquad \text{for } x \in B_\rho, \tag{2.44}$$

since supp $\phi \subset B_r$ and $|x - z| \ge |z|/2$ for $|z| > 2\rho$ (since $|z|/2 + \rho \le |z| \le |x - z| + |x| \le |x - z| + \rho$). Hence

$$\|p_{1,2}\|_{L^{3/2}(B_\rho)} \le c\rho^2 \|p_{1,2}\|_{L^\infty(B_\rho)} \le c\rho^2 \int_{2\rho < |z| < r} \frac{|u(z)|^2}{|z|^3} dz.$$

Note, however, that such estimate is weaker than the required estimate (the last term in (2.42)). In order to obtain the required estimate, we use the following trick: rather than estimating $p_{1,2}$ in (2.44) we estimate $\nabla p_{1,2}$,

$$|\nabla p_{1,2}(x)| \le c \int_{2\rho < |z| < r} \frac{|u(z)|^2 \phi(z)}{|x-z|^4} dz \le c \int_{2\rho < |z| < r} \frac{|u(z)|^2}{|z|^4} dz \qquad \text{for } x \in B_\rho,$$

where we used (2.43) with $|\alpha| = 3$, and use the mean value theorem to write

$$\|p_{1,2} - (p_{1,2})_\rho\|_{L^{3/2}(B_\rho)} \le c\rho^2 \|p_{1,2} - (p_{1,2})_\rho\|_{L^\infty(B_\rho)} \le c\rho^2 \text{diam}(B_\rho) \|\nabla p_{1,2}\|_{L^\infty(B_\rho)}$$

$$\le c\rho^3 \int_{2\rho < |z| < r} \frac{|u(z)|^2}{|z|^4} dz \qquad \text{for } x \in B_\rho,$$

as required. Thus (2.42) follows by writing

$$\|p_1 - (p_1)_\rho\|_{L^{3/2}(B_\rho)} \le \|p_{1,1} - (p_{1,1})_\rho\|_{L^{3/2}(B_\rho)} + \|p_{1,2} - (p_{1,2})_\rho\|_{L^{3/2}(B_\rho)}$$

$$\le 2\|p_{1,1}\|_{L^{3/2}(B_\rho)} + \|p_{1,2} - (p_{1,2})_\rho\|_{L^{3/2}(B_\rho)}$$

$$\le c\|u\|^2_{L^3(B_{2\rho})} + c\rho^3 \int_{2\rho < |y| < r} \frac{|u(z)|^2}{|z|^4} dz,$$

where we used (2.6) in the second inequality.

Step 4. We estimate p_2.

Namely, in this step we will show that

$$\|p_2 - (p_2)_\rho\|_{L^{3/2}(B_\rho)} \leq c \left(\frac{\rho}{r}\right)^3 \left(\|p - (p)_r\|_{L^{3/2}(B_r)} + \|u\|_{L^3(B_r)}^2\right). \tag{2.45}$$

Note this concludes the proof of the lemma (since this together with the estimate on $p_1 - (p_1)_\rho$ above give (2.40)). We will show that

$$\|p_2 - (p_2)_\rho\|_{L^{3/2}(B_\rho)} \leq c \left(\frac{\rho}{r}\right)^3 \|p_2 - (p_2)_r\|_{L^{3/2}(B_r)}. \tag{2.46}$$

Then (2.45) follows by estimating the last $L^{3/2}$ norm,

$$\begin{aligned}
\|p_2 - (p_2)_r\|_{L^{3/2}(B_r)} &\leq \|p - (p)_r\|_{L^{3/2}(B_r)} + \|p_1 - (p_1)_\rho\|_{L^{3/2}(B_r)} \\
&\leq \|p - (p)_r\|_{L^{3/2}(B_r)} + 2\|p_1\|_{L^{3/2}(B_r)} \\
&\leq \|p - (p)_r\|_{L^{3/2}(B_r)} + c\||u|^2\phi\|_{L^{3/2}(\mathbb{R}^3)} \\
&\leq \|p - (p)_r\|_{L^{3/2}(B_r)} + c\|u\|_{L^3(B_r)}^2,
\end{aligned}$$

where we used the triangle inequality, (2.6) and the Calderón–Zygmund inequality (2.39).

As for (2.46), observe that, since p_2 is harmonic in $B_{7r/8}$, the same is true of $p_2 - (p_2)_r$. Thus, since $B(x, r/4) \Subset B_{7r/8}$ for each $x \in B_\rho$, the interior estimate of the first derivative of harmonic functions (see Theorem 7 in Sect. 2.2 of Evans (2010), for example) gives for such x

$$|\nabla p_2(x)| \leq \frac{c}{r^4}\|p_2 - (p_2)_r\|_{L^1(B(x,r/4))} \leq \frac{c}{r^4}\|p_2 - (p_2)_r\|_{L^1(B_r)} \leq \frac{c}{r^3}\|p_2 - (p_2)_r\|_{L^{3/2}(B_r)}$$

where we also used Hölder's inequality. Thus another use of Hölder's inequality and the mean value theorem gives

$$\begin{aligned}
\|p_2 - (p_2)_\rho\|_{L^{3/2}(B_\rho)} &\leq c\rho^2\|p_2 - (p_2)_\rho\|_{L^\infty(B_\rho)} \\
&\leq c\rho^2 \operatorname{diam}(B_\rho)\|\nabla p_2\|_{L^\infty(B_\rho)} \\
&\leq c\frac{\rho^3}{r^3}\|p_2 - (p_2)_r\|_{L^{3/2}(B_r)},
\end{aligned}$$

as required. $\qquad\square$

We now turn to the estimate (2.9).

Lemma 2.10 (The pressure estimate II) *If $p \in L^{3/2}(Q_r)$ satisfies*

$$-\Delta p = \sum_{i,j} \partial_{ij}(u_i u_j) \tag{2.47}$$

in Q_r and $0 < \rho \le r/2$ then

$$P(\rho)^{4/3} \le c \left(\frac{\rho}{r}\right)^{-2} A(r)E(r) + c \left(\frac{\rho}{r}\right)^{4/3} P(r)^{4/3}.$$

The main idea of the lemma is that (2.47) is equivalent to

$$-\Delta p = \sum_{i,j} \partial_{ij}\left(u_i\left(u_j - (u_j)_{B_r}\right)\right), \tag{2.48}$$

since u is divergence-free. Given this observation, the proof is similar to the proof of the previous pressure estimate, which we outline below.

Proof Due to the scale invariance of claim of the lemma, it is sufficient to take $r = 1$. Similarly, as in the previous pressure estimate, we will take $(y, s) = (0, 0)$ and we reduce the claim to an inequality in space only, that is, we will show (in the steps below) that

$$\|p\|_{L^{3/2}(B_\rho)} \le c\|u\|_{L^2(B_1)}\|\nabla u\|_{L^2(B_1)} + c\rho^2 \|p\|_{L^{3/2}(B_1)} \tag{2.49}$$

for each time $t \in (-\rho^2, 0)$. Then taking the 3/2 power, integrating in $\int_{-\rho^2}^0 dt$ we obtain

$$\int_{Q_\rho} |p|^{3/2} \le c \int_{-\rho^2}^0 \|u(t)\|_{L^2(B_1)}^{3/2}\|\nabla u(t)\|_{L^2(B_1)}^{3/2} dt + c\rho^3 \int_{Q_1} |p|^{3/2}$$

$$\le c \left(\int_{-\rho^2}^0 \|u(t)\|_{L^2(B_1)}^6 dt\right)^{1/4} \left(\int_{-\rho^2}^0 \|\nabla u(t)\|_{L^2(B_1)}^2 dt\right)^{3/4} + c\rho^3 \int_{Q_1} |p|^{3/2}$$

$$\le c\rho^{1/2} \left(\sup_{t\in(-1,0)} \|u(t)\|_{L^2(B_1)}^2\right)^{3/4} \left(\int_{-1}^0 \|\nabla u(t)\|_{L^2(B_1)}^2 dt\right)^{3/4} + c\rho^3 \int_{Q_1} |p|^{3/2}$$

$$= c\rho^{1/2} A(1)^{3/4} E(1)^{3/4} + c\rho^3 P(1),$$

where we used Hölder's inequality (in the time integral) in the second line. The claim of the lemma follows by dividing by ρ^2 and taking the 4/3 power.

Step 1. We decompose the pressure function p.
  Namely, let $\phi \in C_0^\infty(B_1; [0, 1])$ be such that $\phi \equiv 1$ on $B_{7/8}$ and

$$p_1 := \sum_{i,j} \partial_{ij}\Psi * \left(u_i\left(u_j - (u_j)_{B_1}\right)\phi\right).$$

Thus letting $p_2 := p - p_1$, we obtain a decomposition of p,

$$p = p_1 + p_2,$$

with p_2 harmonic in $B_{7/8}$.

Step 2. We show that $\|p_1\|_{L^3(B_1)} \leq c\|u\|_{L^2(B_1)}\|\nabla u\|_{L^2(B_1)}$.

We use the Calderón–Zygmund inequality (2.39) together with the Poincaré inequality $\|f - (f)_{B_1}\|_{L^6(B_1)} \leq c\|\nabla f\|_{L^2(B_1)}$ to obtain

$$\|p_1\|_{L^{3/2}(B_\rho)} \leq c \sum_{i,j} \|u_i \left(u_j - (u_j)_{B_1}\right)\|_{L^{3/2}(B_1)}$$

$$\leq c\|u\|_{L^2(B_1)}\|u - (u)_{B_1}\|_{L^6(B_1)}$$

$$\leq c\|u\|_{L^2(B_1)}\|\nabla u\|_{L^2(B_1)}.$$

Step 3. We show that $\|p_2\|_{L^{3/2}(B_\rho)} \leq c\rho^2 \|p\|_{L^{3/2}(B_1)} + \|p_1\|_{L^3(B_1)}$. (Note this and the previous step prove (2.49), as required.)

Since $B_{1/4}(x) \subset B_{7/8}$ for every $x \in B_\rho$, we can use the mean value property of harmonic functions to obtain

$$\int_{B_\rho} |p_2|^{3/2} = \int_{B_\rho} \left| \frac{1}{|B_{1/4}|} \int_{B_{1/4}(x)} p_2 \right|^{3/2} dx$$

$$\leq \int_{B_\rho} \left(\frac{1}{|B_{1/4}|} \int_{B_{1/4}(x)} |p_2|^{3/2} \right) dx$$

$$\leq c\rho^3 \int_{B_1} |p_2|^{3/2}.$$

Using the triangle inequality (and the fact that $\rho < 1$), we obtain the claim of this step.   □

Chapter 3
Weak Solution of the Navier–Stokes Inequality with a Point Blow-Up

In this chapter, we construct the first of Scheffer's counterexamples that is a weak solution of the Navier–Stokes inequality that blows up in finite time at a point. Namely, we prove Theorem 1.9, which we restate for the reader's convenience.

Theorem 3.1 (Weak solution of NSI with point singularity) *There exist $\nu_0 > 0$ and a function $u\colon \mathbb{R}^3 \times [0, \infty) \to \mathbb{R}^3$ that is a weak solution of the Navier–Stokes inequality for all $\nu \in [0, \nu_0]$ such that $u(t) \in C^\infty$, $\mathrm{supp}\, u(t) \subset G$ for all t for some compact set $G \Subset \mathbb{R}$ (independent of t). Moreover, u is unbounded in every neighbourhood of (x_0, T_0), for some $x_0 \in \mathbb{R}^3$, $T_0 > 0$.*

This theorem was first proved by Scheffer (1985). In the next chapter, we extend this result to the case when the blow-up occurs on a Cantor set S with $\xi \le d_H(S) \le 1$ for any preassigned $\xi \in (0, 1)$, rather than a one-element singular set $S = \{(x_0, T_0)\}$.

Recall that, using an appropriate rescaling, the statement of the theorem is equivalent to the one where $\nu = 1$ and $(x_0, T_0) = (0, 1)$. We also recall that a pair (u, p) is a weak solution of the NSI on $\mathbb{R}^3 \times (0, \infty)$ if, according to Definition 1.6, $u \in L^\infty((0, \infty); L^2(\mathbb{R}^3))$, $\nabla u \in L^2(\mathbb{R}^3 \times (0, \infty))$, $u(t)$ is divergence free for almost every $t > 0$, $p \in L^{3/2}_{loc}(\mathbb{R}^3 \times (0, \infty))$ (**the regularity properties**), the equation $-\Delta p = \sum_{i,j=1}^3 \partial_i \partial_j (u_i u_j)$ holds in the sense of distributions in $\mathbb{R}^3$ for almost every $t \in (a, b)$ (**the relation between u and** p), and

$$\int_{\mathbb{R}^3} |u(t)|^2 \phi(t) + 2\nu \int_a^t \int_{\mathbb{R}^3} |\nabla u|^2 \phi \le \int_a^t \int_{\mathbb{R}^3} \left(|u|^2 (\partial_t \phi + \nu \Delta \phi) + (|u|^2 + 2p)(u \cdot \nabla)\phi \right) \tag{3.1}$$

is valid for every $\phi \in C_0^\infty(\mathbb{R}^3 \times (a, b); [0, \infty))$ and $t \in (a, b)$ (**the local energy inequality**). In what follows, we will consider only pairs (u, p) where p is given by

$$p(t) := \sum_{i,j=1}^3 \partial_{ij} \Psi * (u_i(t) u_j(t)), \tag{3.2}$$

© Springer Nature Switzerland AG 2019
W. S. Ożański, *The Partial Regularity Theory of Caffarelli, Kohn, and Nirenberg and its Sharpness*, Advances in Mathematical Fluid Mechanics, https://doi.org/10.1007/978-3-030-26661-5_3

for every t, which we shall refer to simply as the *pressure function correspond-ing to u*. (Recall that $\Psi(x) := (4\pi|x|)^{-1}$ denotes the fundamental solution of the Laplace equation in $\mathbb{R}^3$.) Note that such p immediately satisfies the relation $-\Delta p = \sum_{i,j=1}^{3} \partial_i \partial_j (u_i u_j)$ as well as the required regularity (i.e. $p \in L_{loc}^{3/2}$) if u does. (In fact, due to the Calderón–Zygmund inequality (2.39), we have a stronger property that $p \in L^{5/3}(\mathbb{R}^3 \times (0, \infty))$, recall (1.7)).) Thus, a divergence-free vector field (together with the corresponding pressure function) is a weak solution of the NSI on $\mathbb{R}^3 \times (0, \infty)$ if $u \in L^\infty((0, \infty); L^2(\mathbb{R}^3))$, $\nabla u \in L^2(\mathbb{R}^3 \times (0, \infty))$ and the local energy inequality (3.1) holds.

The proof of the above theorem makes use of an alternative form of the local energy inequality (3.1). Namely, (3.1) is satisfied if the *local energy inequality on the time interval* $[S, S']$,

$$\int_{\mathbb{R}^3} |u(x, S')|^2 \phi \, dx - \int_{\mathbb{R}^3} |u(x, S)|^2 \phi \, dx + 2\nu \int_S^{S'} \int_{\mathbb{R}^3} |\nabla u|^2 \phi$$
$$\leq \int_S^{S'} \int_{\mathbb{R}^3} \left(|u|^2 + 2p \right) u \cdot \nabla \phi + \int_S^{S'} \int_{\mathbb{R}^3} |u|^2 \left(\partial_t \phi + \nu \Delta \phi \right),$$

$$(3.3)$$

holds for all $S, S' > 0$ with $S < S'$, which is clear by taking S, S' such that supp $\phi \subset \mathbb{R}^3 \times (S, S')$. An advantage of this alternative form of the local energy inequality is that it demonstrates how to combine solutions of the Navier–Stokes inequality one after another. Namely, (3.3) shows that a necessary and sufficient condition for two vector fields $u^{(1)}: \mathbb{R}^3 \times [t_0, t_1] \to \mathbb{R}^3$, $u^{(2)}: \mathbb{R}^3 \times [t_1, t_2] \to \mathbb{R}^3$ which satisfy the local energy inequality on the time intervals $[t_0, t_1]$, $[t_1, t_2]$, respectively, to combine (one after another) into a vector field satisfying the local energy inequality on the time interval $[t_0, t_2]$ is that

$$|u^{(2)}(x, t_1)| \leq |u^{(1)}(x, t_1)| \qquad \text{for a.e. } x \in \mathbb{R}^3. \qquad (3.4)$$

Given the above observation, the main idea of the proof of Theorem 3.1 is a certain switching procedure, which we discuss in the following section.

3.1 Sketch of the Proof of Theorem 3.1

Here we present a simple argument which proves Theorem 3.1 given the following assumptions. Namely, suppose for a moment that there exists $T > 0$, a compact set $G \subset \mathbb{R}^3$ and a divergence-free vector field u such that $u \in C^\infty(\mathbb{R}^3 \times [0, T]; \mathbb{R}^3)$, supp $u(t) = G$ for all $t \in [0, T]$, and the Navier–Stokes inequality

$$\partial_t |u|^2 \leq -u \cdot \nabla \left(|u|^2 + 2p \right) + 2\nu u \cdot \Delta u \qquad (3.5)$$

holds in $\mathbb{R}^3 \times [0, T]$ for all $\nu \in [0, \nu_0]$ for some $\nu_0 > 0$, where $p(t)$ is the pressure function corresponding to u (recall (3.2)). Here $C^\infty(\mathbb{R}^3 \times [0, T]; \mathbb{R}^3)$ is a shorthand notation for the space of vector functions that are infinitely differentiable on $\mathbb{R}^3 \times (-\eta, T + \eta)$ for some $\eta > 0$.

Suppose further that, during time interval $[0, T]$ u admits the following interior gain of magnitude property: that for some $\tau \in (0, 1)$, $z \in \mathbb{R}^3$ the affine map

$$\Gamma(x) := \tau x + z,$$

maps G into itself and that, at time T, u attains a large gain in magnitude, namely, that

$$|u(\Gamma(x), T)| \geq \tau^{-1} |u(x, 0)|, \qquad x \in \mathbb{R}^3. \tag{3.6}$$

Such a gain in magnitude allows us to consider a rescaled copy of u and, in a sense, slot it into the part of the support G in which the gain occurred. Namely, considering

$$u^{(1)}(x, t) := \tau^{-1} u(\Gamma^{-1}(x), \tau^{-2}(t - T))$$

we see that $u^{(1)}$ satisfies the Navier–Stokes inequality (3.5) on $\mathbb{R}^3 \times [T, (1 + \tau^2)T]$, $\operatorname{supp} u^{(1)}(t) = \Gamma(G)$ for all $t \in [T, (1 + \tau^2)T]$ and that (3.6) gives

$$\left| u^{(1)}(x, T) \right| \leq |u(x, T)|, \qquad x \in \mathbb{R}^3 \tag{3.7}$$

(and so u, $u^{(1)}$ can be combined "one after another", recall (3.4) above). Thus, since $u^{(1)}$ is larger in magnitude than u (by the factor of τ) and its time of existence is $[T, (1 + \tau^2)T]$, we see that by iterating such a switching we can obtain a vector field $\mathfrak{u}$ that grows indefinitely in magnitude, while its support shrinks to a point (and thus will satisfy all the claims of Theorem 3.1), see Fig. 3.1. To be more precise, we let $t_0 := 0$,

$$t_j := T \sum_{k=0}^{j-1} \tau^{2k} \qquad \text{for } j \geq 1,$$

$T_0 := \lim_{j \to \infty} t_j = T/(1 - \tau^2)$, $u^{(0)} := u$, and

$$u^{(j)}(x, t) := \tau^{-j} u\left(\Gamma^{-j}(x), \tau^{-2j}(t - t_j) \right), \qquad j \geq 1, \tag{3.8}$$

see Fig. 3.1. Clearly

$$\operatorname{supp} u^{(j)}(t) = \Gamma^j(G) \qquad \text{for } t \in [t_j, t_{j+1}] \tag{3.9}$$

and, as in (3.7), (3.6) gives that the magnitude of the consecutive vector fields shrinks at every switching time, that is,

$$\left| u^{(j)}(x, t_j) \right| \leq \left| u^{(j-1)}(x, t_j) \right|, \qquad x \in \mathbb{R}^3, j \geq 1, \tag{3.10}$$

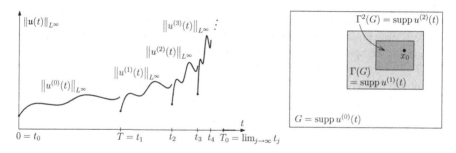

Fig. 3.1 The switching procedure: the blow-up of $\|u(t)\|_\infty$ (left) and the shrinking support of $u(t)$ (right) as $t \to T_0^-$

see Fig. 3.1.

Thus letting

$$u(t) := \begin{cases} u^{(j)}(t) & \text{if } t \in [t_j, t_{j+1}) \text{ for some } j \geq 0, \\ 0 & \text{if } t \geq T_0, \end{cases}$$

we obtain a vector field that satisfies the claims of Theorem 3.1. Indeed, by construction, u is divergence free, smooth in space, its support in space is contained in G, and u is unbounded in every neighbourhood of (x_0, T_0), where

$$\{x_0\} := \bigcap_{j \geq 0} \Gamma^j(G) = \left\{ \frac{z}{1-\tau} \right\}.$$

As for the regularity $u \in L^\infty((0, \infty); L^2(\mathbb{R}^3))$ and $\nabla u \in L^2(\mathbb{R}^3 \times (0, \infty))$ (recall Definition 1.6), we write for any $t \in [t_j, t_{j+1}], j \geq 0$,

$$\|u(t)\| = \|u^{(j)}(t)\| \leq \sup_{t \in [t_j, t_{j+1})} \|u^{(j)}(t)\| = \tau^{j/2} \sup_{t \in [t_0, t_1]} \|u^{(0)}(t)\| \leq \sup_{t \in [t_0, t_1]} \|u^{(0)}(t)\| < \infty,$$

$$(3.11)$$

where we used the fact that $\tau \in (0, 1)$ and we used the shorthand notation $\|\cdot\| \equiv \|\cdot\|_{L^2(\mathbb{R}^3)}$. Similarly,

$$\int_0^\infty \|\nabla u(t)\|^2 = \sum_{j=0}^\infty \int_{t_j}^{t_{j+1}} \|\nabla u^{(j)}(t)\|^2 = \int_{t_0}^{t_1} \|\nabla u^{(0)}(t)\|^2 \sum_{j=0}^\infty \tau^j < \infty, \quad (3.12)$$

as required.

As for the local energy inequality (3.1), we see that, by construction, the local energy inequality (3.3) is satisfied on any time interval $[S, S'] \subset [0, T_0)$. Since $\|u(t)\| \to 0$ as $t \to T_0^-$ (since $\tau^{2j} \to 0$ as $j \to \infty$, see the calculation above) and the regularity $\sup_{t>0} \|u(t)\| < \infty$, $\nabla u \in L^2(\mathbb{R}^3 \times (0, \infty))$) gives global-in-time integrability of all the terms appearing under the space–time integrals in (3.3) the Dominated Convergence Theorem lets us take the limit $S' \to T_0$ to obtain the local energy inequality on any interval $[S, S'] \subset [0, \infty)$, as required.

Therefore, we have established the proof of Theorem 3.1 given the existence of T, G, u, ν_0, τ and z with the properties listed above. These objects are constructed in Sect. 3.3 (which includes a particularly enlightening proof of the Navier–Stokes inequality (3.5), see Sect. 3.3.2). We now discuss some interesting properties of the vector field u which are consequences of the above switching procedure.

3.1.1 Remarks

Note that u enjoys a self-similar property

$$u(x_0 - x, T_0 - s) = \tau^j u(x_0 - \tau^j x, T_0 - \tau^{2j} s), \qquad x \in \mathbb{R}^3, s \in (0, T_0], j \geq 0,$$

which is also the property characteristic for the Leray hypothetical self-focusing strong solutions to the Navier–Stokes equations (that is, (3.12) in Leray (1934), in which $x_0 = 0$; note, however, such solutions do not exist, as was shown by Nečas, Růžička & Šverák, 1996), except that here the self-similarity holds only for the discrete scaling factors τ^j, $j \geq 0$.

Moreover, u satisfies the energy inequality

$$\|u(\tau_2)\|_{L^2}^2 + 2\nu \int_{\tau_1}^{\tau_2} \|\nabla u(t)\|_{L^2}^2 dt \leq \|u(\tau_1)\|_{L^2}^2, \qquad \nu \in [0, \nu_0] \qquad (3.13)$$

for every $\tau_1 \in [0, \infty)$ such that $\tau_1 \notin \{t_j\}_{j \geq 1}$, and every $\tau_2 > \tau_1$ (where we used the shorthand notation $L^2 \equiv L^2(\mathbb{R}^3)$), which can be verified as follows. Let $1 \leq j_1 \leq j_2$ and take

$$\phi(x, t) = \psi(x) \mathcal{T}(t),$$

where $\psi \in C_0^\infty(\mathbb{R}^3)$ is such that $\psi \geq 0$, $\psi = 1$ on G and $\mathcal{T} \in C_0^\infty((0, \infty))$ is such that $\mathcal{T} = 1$ on $[t_{j_1}, t_{j_2}]$ and $\operatorname{supp} \mathcal{T} \subset (t_{j_1-1}, t_{j_2+1})$. Then the local energy inequality (3.3) and the fact that $\operatorname{supp} u(t) \subset G$ give

$$2\nu \int_{t_{j_1-1}}^{t_{j_2+1}} \mathcal{T}(t) \|\nabla u(t)\|_{L^2}^2 dt \leq \int_{t_{j_1-1}}^{t_{j_1}} \|u(t)\|_{L^2}^2 \mathcal{T}'(t)\, dt + \int_{t_{j_2}}^{t_{j_2+1}} \|u(t)\|_{L^2}^2 \mathcal{T}'(t)\, dt.$$

Given $\varepsilon > 0$ and $\tau_1 \in (t_{j_1-1}, t_{j_1})$, $\tau_2 \in (t_{j_2}, t_{j_2+1})$ let $\mathcal{T}(t) := J_\varepsilon \chi_{(\tau_1, \tau_2)}(t)$, where χ is an indicator function and J_ε denotes the (usual) mollification operator. Given such a choice of $\mathcal{T}$ we can use the smoothness of u on each of the intervals (t_j, t_{j+1}), $j \geq 0$ to take the limit $\varepsilon \to 0^+$ in the inequality above to obtain the energy inequality (3.13) for $\tau_1 \in (t_{j_1-1}, t_{j_1})$, $\tau_2 \in (t_{j_2}, t_{j_2+1})$. Thus, since u is right-continuous in time and its magnitude does not increase at a switching time (recall (3.10)), the last inequality is valid also for $\tau_1 \in [t_{j_1-1}, t_{j_1})$, $\tau_2 \in [t_{j_2}, t_{j_2+1}]$, as required.

Furthermore, although the vector field u is not a solution of the Navier–Stokes equations, it can be used to benchmark some results in the theory of these equations, for example, the regularity criteria. A regularity criterion is a condition guaranteeing that a local-in-time strong solution u of the Navier–Stokes equations on a time interval $[0, T)$ does not blow-up as $t \to T^-$. For example, $u(t)$ does not blow-up if it satisfies any of the following:

(1) The *Beale–Kato–Majda criterion* (due to Beale, Kato & Majda, 1984):

$$\int_0^T \|\text{curl } u(t)\|_{L^\infty} < \infty,$$

(2) The *Serrin condition* (due to Serrin 1963):

$$\int_0^T \|u(t)\|_{L^s}^r < \infty \qquad \text{for any } s \geq 3, r \geq 2 \text{ satisfying } \frac{2}{r} + \frac{3}{s} = 1,$$

or

(3) *Control of the direction of vorticity* (due to Constantin & Feffeman, 1993):

$$\text{for some } \Omega, \rho > 0 \quad \left|P_{\xi(x,t)}^\perp(\xi(x+y, t))\right| \leq |y|/\rho \qquad (3.14)$$

for x, y, t such that

$$t \in [0, T], \quad |\text{curl } u(x, t)|, |\text{curl } u(x+y, t)| > \Omega.$$

Here $\xi(x, t) := \text{curl } u(x, t)/|\text{curl } u(x, t)|$ is the direction of vorticity curl $u(x, t)$, and $P_x^\perp y := \sin \alpha$, where α denotes the angle between the vectors $x, y \in \partial B(0, 1) \subset \mathbb{R}^3$.

Remarkably, u does not satisfy any of the above criteria, which is a consequence of the switching argument applied in the previous section (as for (3) above note that the direction of curl $u^{(0)}$ is not constant and so the direction of $u^{(j)}$ cannot be controlled as in (3.14) as $j \to \infty$).

However, u does satisfy the $L_{3,\infty}$ criterion (due to Constantin & Feffeman, 2003, see also Seregin, 2007, 2012): if

$$\|u(t)\|_{L^3(\mathbb{R}^3)} \text{ remains bounded as } t \to T^-$$

then $u(t)$ (a local-in-time strong solution on time interval $[0, T)$) does not blow-up as $t \to T^-$. Indeed the L^3 norm of $u(t)$ remains bounded by $\sup_{t\in[0,T]} \|u^{(0)}(t)\|_{L^3(\mathbb{R}^3)}$. This shows that the $L_{3,\infty}$ regularity criterion uses, in an essential way, properties of solutions of the Navier–Stokes equations (rather than merely the Navier–Stokes inequality (1.16)).

In the next three sections, we complete the sketch of the proof of Theorem 3.1, that is, we construct constants $T > 0$, $\nu_0 > 0$, $\tau \in (0, 1)$, $z \in \mathbb{R}^3$, the set G and the vector field u with the properties listed in the beginning of Sect. 3.1. For this we first introduce a number of preliminary results regarding axisymmetric vector fields in $\mathbb{R}^3$, properties of the pressure function as well as introduce the concept of a *structure* on a subset U of the upper half-plane (Sect. 3.2). Then, in Sect. 3.3, we perform the construction of $T > 0$, $\nu_0 > 0$, $\tau \in (0, 1)$, $z \in \mathbb{R}^3$, G, u and we show the required claims. The construction is based on a certain *geometric arrangement*, which is the heart of the proof of Theorem 3.1 and which we discuss in detail in Sect. 3.4.

3.2 Preliminaries

We will say that a function is *smooth* on an open set if it is of class C^∞ on this set. We use the notation ∂_λ for the partial derivative with respect to a variable λ. We often simplify the notation corresponding to the partial derivative with respect to x_i by writing

$$\partial_i \equiv \partial_{x_i}.$$

We do not apply the summation convention over repeated indices. We let

$$P := \{(x_1, x_2) \in \mathbb{R}^2 \colon x_2 > 0\}$$

denote the upper half-plane. We frequently use the convention

$$h_t(\cdot) \equiv h(\cdot, t), \tag{3.15}$$

that is, the subscript t denotes dependence on t (rather than the t-derivative, which we denote by ∂_t). By writing

"*outside G*"   we mean   "*for $x \notin G$*".

By $\overline{U}$ we denote the closure of an open set U. We often write that a function is *a solution of* a theorem (or proposition/lemma) if it satisfies the claim of the theorem.

3.2.1 The Rotation R_φ

We denote by R_φ the rotation around the x_1 axis by an angle φ, that is,

$$R_\varphi(x_1, x_2, x_3) = (x_1, x_2 \cos \varphi - x_3 \sin \varphi, x_2 \sin \varphi + x_3 \cos \varphi).$$

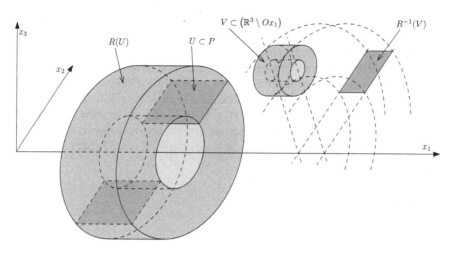

Fig. 3.2 The rotation R and the cylindrical projection R^{-1}

We will refer to R_φ (for some φ) simply as *the rotation* since it is the only operation of rotation that we will consider. It is clear that any $x \in \mathbb{R}^3$ is either a point on the x_1 axis, a point in P or a rotation $R_\varphi(y_1, y_2, 0)$ of some $y \in P$ by some angle $\varphi \in (0, 2\pi)$. For $U \Subset P$ set

$$R(U) := \{x \in \mathbb{R}^3 \ : \ x = R_\varphi(y, 0) \text{ for some } \varphi \in [0, 2\pi), \ y \in U\}, \qquad (3.16)$$

the *rotation of U* (see Fig. 3.2). Clearly, if U_1, U_2 are disjoint subsets of P then $R(U_1)$, $R(U_2)$ are disjoint subsets of $\mathbb{R}^3$. We will denote by $R^{-1} \colon \mathbb{R}^3 \to \overline{P}$ the *cylindrical projection* defined by

$$R^{-1}(y_1, y_2, y_3) := \left(y_1, \sqrt{y_2^2 + y_3^2}\right). \qquad (3.17)$$

The projection R^{-1} is in fact the left-inverse of R, that is, $R^{-1}R = \text{id}$. It is not a right-inverse, but $R R^{-1}(V) \supset V$ for any $V \subset \left(\mathbb{R}^3 \setminus Ox_1\right)$ (where Ox_1 denotes the x_1 axis), as is clear from Fig. 3.2. We say that a velocity field $u \colon \mathbb{R}^3 \to \mathbb{R}^3$ is *axisymmetric* if

$$u(R_\varphi x) = R_\varphi u(x) \qquad \text{for } \varphi \in [0, 2\pi), x \in \mathbb{R}^3, \qquad (3.18)$$

while a scalar function $q \colon \mathbb{R}^3 \to \mathbb{R}$ is *axisymmetric* if

$$q(R_\varphi x) = q(x) \qquad \text{for } \varphi \in [0, 2\pi), x \in \mathbb{R}^3;$$

in other words, $q(x) = q(R^{-1}x)$ for $x \in \mathbb{R}^3$. Observe that if a vector field $u \in C^2$ and a scalar function $q \in C^1$ are axisymmetric then the vector function $(u \cdot \nabla)u$ and the scalar functions

$$|u|^2, \quad \text{div}\, u, \quad u \cdot \nabla |u|^2, \quad u \cdot \nabla q, \quad u \cdot \Delta u \quad \text{and} \quad \sum_{i,j=1}^{3} \partial_i u_j \partial_j u_i \qquad (3.19)$$

are axisymmetric. These facts can be verified by a simple calculation and by making use of the algebraic identity

$$\sum_{i,j=1}^{3} \partial_i u_j \partial_j u_i = \text{div}\left((u \cdot \nabla)u\right) - u \cdot \nabla(\text{div}\, u),$$

see Sect. 3.6 for details.

3.2.2 The Pressure Function

Given a vector field $u \colon \mathbb{R}^3 \to \mathbb{R}^3$ consider the pressure function $p \colon \mathbb{R}^3 \to \mathbb{R}$ corresponding to u, that is,

$$p(x) := \int_{\mathbb{R}^3} \sum_{i,j=1}^{3} \frac{\partial_i u_j(y) \partial_j u_i(y)}{4\pi |x - y|}\, dy,$$

recall (3.2). Here we briefly comment on some geometric properties of the pressure function, which will be crucial in constructing a velocity field u satisfying the Navier–Stokes inequality (3.5) (see, for instance, Lemma 3.6).

First, if $u \in C_0^\infty(\mathbb{R}^3)$ then the corresponding pressure function is smooth on $\mathbb{R}^3$ with

$$|\nabla p(x)| \le \tilde{C}|x|^{-4} \quad \text{and} \quad |D^2 p(x)| \le \tilde{C}|x|^{-5} \qquad (3.20)$$

for some $\tilde{C} > 0$ (which depends on u), which follows from integration by parts. Moreover, p satisfies the limiting property

$$\lim_{x_1 \to \pm\infty} x_1^4 \partial_1 p(x_1, 0, 0) = \frac{\pm 3}{4\pi} \int_{\mathbb{R}^3} \left(|u(y)|^2 - 3u_1^2(y)\right) dy, \qquad (3.21)$$

which can be verified directly. Finally, if $u \in C_0^\infty(\mathbb{R}^3)$ is axisymmetric then the change of variable $z = R_{-\varphi} y$ and (3.19) give

$$p(R_\varphi x) = \int_{\mathbb{R}^3} \sum_{i,j=1}^3 \frac{\partial_i u_j(y) \partial_j u_i(y)}{4\pi |R_\varphi x - y|} dy = \int_{\mathbb{R}^3} \sum_{i,j=1}^3 \frac{\partial_i u_j(R_\varphi z) \partial_j u_i(R_\varphi z)}{4\pi |x - z|} dz = p(x)$$

(3.22)

for all $\varphi \in [0, 2\pi)$. That is, the pressure function corresponding to an axisymmetric vector field is axisymmetric.

3.2.3 The Functions $u[v, f]$, $p[v, f]$

Now let v be a 2D vector field and f be a scalar function defined on P such that

$$v \in C_0^\infty(P; \mathbb{R}^2), \ f \in C_0^\infty(P; [0, \infty)), \text{ and } f > |v| \text{ on supp } v. \quad (3.23)$$

For such v, f we define $u[v, f] \colon \mathbb{R}^3 \to \mathbb{R}^3$ to be the axisymmetric vector field satisfying

$$u[v, f](x_1, x_2, 0) := \left(v_1(x_1, x_2), v_2(x_1, x_2), \sqrt{f(x_1, x_2)^2 - |v(x_1, x_2)|^2} \right) \quad (3.24)$$

for $(x_1, x_2) \in \overline{P}$. (Here $\overline{P}$ denotes the closure of P.) Note that such definition immediately gives

$$|u[v, f]| = f. \quad (3.25)$$

We emphasise that (3.23) also implies that supp $u[v, f]$ is strictly separated from the Ox_1 axis (the rotation axis).

Moreover, the definition can be rewritten in a simple, equivalent form using cylindrical coordinates x_1, ρ, φ. Namely,

$$u[v, f](x_1, \rho, \varphi) = v_1(x_1, \rho)\widehat{x_1} + v_2(x_1, \rho)\widehat{\rho} + \sqrt{f(x_1, \rho)^2 - |v(x_1, \rho)|^2}\,\widehat{\varphi},$$

(3.26)

where the cylindrical coordinates are defined using the representation

$$\begin{cases} x_1 = x_1, \\ x_2 = \rho \cos \varphi, \\ x_3 = \rho \sin \varphi \end{cases}$$

and the cylindrical unit vectors $\widehat{x_1}, \widehat{\rho}, \widehat{\varphi}$ are

$$\begin{cases} \widehat{x_1}(x_1, \rho, \varphi) := (1, 0, 0), \\ \widehat{\rho}(x_1, \rho, \varphi) := (0, \cos \varphi, \sin \varphi), \\ \widehat{\varphi}(x_1, \rho, \varphi) := (0, -\sin \varphi, \cos \varphi). \end{cases} \quad (3.27)$$

In particular, for this coordinate system the chain rule gives

$$\partial_\rho = \cos\varphi\,\partial_{x_2} + \sin\varphi\,\partial_{x_3},$$
$$\partial_\varphi = -\rho\sin\varphi\,\partial_{x_2} + \rho\cos\varphi\,\partial_{x_3}. \tag{3.28}$$

Clearly, if supp v, supp $q \subset U$ for some $U \subset P$ then supp $u[v, q] \subset R(U)$. Moreover, since both v and f have compact support in P and since $f > |v|$ on supp v (so that $\sqrt{f^2 - |v|^2} \in C_0^\infty(P)$) it is clear that $u[v, f] \in C_0^\infty(\mathbb{R}^3; \mathbb{R}^3)$. The vector field $u[v, f]$ enjoys some further useful properties, which we have shown below.

Lemma 3.2 (Properties of $u[v, f]$)

(i) *The vector field $u[v, f]$ is divergence free if and only if v satisfies*

$$\mathrm{div}(x_2\,v(x_1, x_2)) = 0 \qquad \text{for all } (x_1, x_2) \in P.$$

(ii) *If $v \equiv 0$ then*

$$\Delta u[0, f](x_1, \rho, \varphi) = Lf(x_1, \rho)\widehat{\varphi},$$

where

$$Lf(x_1, x_2) := \Delta f(x_1, x_2) + \frac{1}{x_2}\partial_{x_2}f(x_1, x_2) - \frac{1}{x_2^2}f(x_1, x_2). \tag{3.29}$$

In particular

$$\Delta u[0, f](x_1, x_2, 0) = (0, 0, Lf(x_1, x_2)). \tag{3.30}$$

(iii) *For all $x_1, x_2 \in \mathbb{R}$*

$$\partial_{x_3}|u[v, f]|(x_1, x_2, 0) = 0. \tag{3.31}$$

Proof The lemma is a consequence of elementary calculations using cylindrical coordinates, which we now briefly discuss.

As for (i) recall that the divergence of a vector field u described in cylindrical coordinates as $u = u_1\widehat{x_1} + u_\rho\widehat{\rho} + u_\varphi\widehat{\varphi}$ is

$$\mathrm{div}\,u = \partial_{x_1}u_1 + \frac{1}{\rho}\partial_\rho\left(\rho u_\rho\right) + \frac{1}{\rho}\partial_\varphi u_\varphi.$$

Thus, since $u[v, f]_\varphi = \sqrt{f^2 - |v|^2}$ does not depend on φ we obtain (i).

As for (ii) recall that the Laplacian of any function $F = F(x_1, \rho, \varphi)$ is

$$\Delta F = \partial_{x_1 x_1}F + \frac{1}{\rho}\partial_\rho\left(\rho\,\partial_\rho F\right) + \frac{1}{\rho^2}\partial_{\varphi\varphi}F.$$

Thus, since $u[0, f] = f\widehat{\varphi}$ and because the unit vector $\widehat{\varphi}$ depends only on φ and satisfies $\partial_{\varphi\varphi}\widehat{\varphi} = -\widehat{\varphi}$ (recall (3.27)) we obtain

$$\Delta u[0, f] = \partial_{x_1 x_1} f(x_1, \rho)\widehat{\varphi} + \frac{1}{\rho}\partial_\rho \left(\rho\, \partial_\rho f(x_1, \varphi)\right)\widehat{\varphi} + \frac{f(x_1, \rho)}{\rho^2}\partial_{\varphi\varphi}\widehat{\varphi}$$

$$= \partial_{x_1 x_1} f(x_1, \rho)\widehat{\varphi} + \frac{1}{\rho}\partial_\rho f(x_1, \varphi)\widehat{\varphi} + \partial_{\rho\rho} f(x_1, \varphi)\widehat{\varphi} - \frac{f(x_1, \rho)}{\rho^2}\widehat{\varphi}$$

$$= Lf(x_1, \rho)\widehat{\varphi}.$$

In particular, taking $\varphi = 0$ gives (3.30).

As for (iii) it is enough to note that since $|u[v, f]| = f(x_1, \rho)$ is axisymmetric, the derivative in question is in fact a derivative along a level set of $|u[v, f]|$ (that is, along a circle around the x_1 axis). In other words, the relations (3.28) give

$$\partial_{x_3} = \sin\varphi\, \partial_\rho + \frac{\cos\varphi}{\rho}\partial_\varphi \qquad (3.32)$$

and so, because $|u[v, f]| = f$ does not depend on φ,

$$\partial_{x_3}|u[v, f]| = \sin\varphi\left(\partial_\rho f\right),$$

which vanishes when $\varphi = 0, \pi$. $\qquad\qquad\square$

We define $p^*[v, f]: \mathbb{R}^3 \to \mathbb{R}$ to be the pressure function corresponding to $u[v, f]$, that is,

$$p^*[v, f](x) := \int_{\mathbb{R}^3} \sum_{i,j=1}^{3} \frac{\partial_i u_j[v, f](y)\partial_j u_i[v, f](y)}{4\pi|x - y|} dy, \qquad (3.33)$$

and we denote its restriction to $\mathbb{R}^2$ by $p[v, f]$, that is,

$$p[v, f](x_1, x_2) := p^*[v, f](x_1, x_2, 0). \qquad (3.34)$$

It is clear that, since $u[v, f] \in C_0^\infty(\mathbb{R}^3)$,

$$p[v, f] \in C^\infty(\mathbb{R}^2). \qquad (3.35)$$

Furthermore, since $u[v, f]$ is axisymmetric, the same is true of $p^*[v, f]$ (recall (3.22)). In particular, in the same way as in the proof of Lemma 3.2 (iii) above we obtain that

$$\partial_{x_3} p^*[v, f](x_1, x_2, 0) = 0 \quad \text{for all } x_1, x_2 \in \mathbb{R}. \qquad (3.36)$$

Similarly,

$$\partial_{x_2} p^*[v, f](x_1, 0, x_3) = 0 \quad \text{for all } x_1, x_3 \in \mathbb{R}, \qquad (3.37)$$

using the relation

$$\partial_{x_2} = \cos\varphi\, \partial_\rho - \frac{\sin\varphi}{\rho}\partial_\varphi, \qquad (3.38)$$

which is a consequence of (3.28). Thus, taking $x_3 = 0$ in (3.37) we obtain

$$\partial_{x_2} p[v, f](x_1, 0) = 0 \quad \text{for } x_1 \in \mathbb{R}. \tag{3.39}$$

The function $p[v, f]$ enjoys some further properties, which we state in a lemma.

Lemma 3.3 (Properties of $p[v, f]$) *Let $v = (v_1, v_2)$, f be as in (3.23). Then*

(i) $p[v, f] = p[-v, f]$,

(ii) *if additionally $v_2(\cdot, x_2)$ is odd and $v_1(\cdot, x_2)$, $f(\cdot, x_2)$ are even for each fixed x_2 then $p[v, f]$ is even, that is,*

$$p[v, f](x) = p[v, f](-x) \quad \text{for all } x \in \mathbb{R}^2,$$

(iii) *if $\tilde{v}$, $\tilde{f}$ is another pair satisfying (3.23) and such that f, $\tilde{f}$ have disjoint supports then*

$$p\left[v + \tilde{v}, f + \tilde{f}\right] = p[v, f] + p\left[\tilde{v}, \tilde{f}\right].$$

Proof Property (iii) follows directly from definition (3.33). As for (i), we will show that $p^*[v, f] = p^*[-v, f]$. Substituting (3.27) into (3.26) we obtain

$$u_1[v, f](x_1, \rho, \varphi) = v_1(x_1, \rho),$$
$$u_2[v, f](x_1, \rho, \varphi) = v_2(x_1, \rho) \cos \varphi - \sqrt{f^2 - |v|^2}(x_1, \rho) \sin \varphi, \tag{3.40}$$
$$u_3[v, f](x_1, \rho, \varphi) = v_2(x_1, \rho) \sin \varphi + \sqrt{f^2 - |v|^2}(x_1, \rho) \cos \varphi.$$

Thus, since for $\varphi = 0$ we have $\partial_2 = \partial_\rho$, $\partial_3 = \rho^{-1} \partial_\varphi$ (see (3.32), (3.38)) and we obtain

$$
\begin{array}{lll}
\partial_1 u_1[v, f] = \partial_{x_1} v_1, & \partial_2 u_1[v, f] = \partial_\rho v_1, & \partial_3 u_1[v, f] = 0, \\
\partial_1 u_2[v, f] = \partial_{x_1} v_2, & \partial_2 u_2[v, f] = \partial_\rho v_2, & \partial_3 u_2[v, f] = -\sqrt{f^2 - |v|^2}/\rho, \\
\partial_1 u_3[v, f] = \partial_{x_1} \sqrt{f^2 - |v|^2}, & \partial_2 u_3[v, f] = \partial_\rho \sqrt{f^2 - |v|^2}, & \partial_3 u_3[v, f] = v_2/\rho,
\end{array}
$$
$$\tag{3.41}$$

from which we immediately see that

$$\partial_i u_j[v, f] \partial_j u_i[v, f] = \partial_i u_j[-v, f] \partial_j u_i[-v, f]$$

for any choice of $i, j \in \{1, 2, 3\}$. Summation in i, j gives

$$\sum_{i,j=1}^{3} \partial_i u_j[v, f] \partial_j u_i[v, f] = \sum_{i,j=1}^{3} \partial_i u_j[-v, f] \partial_j u_i[-v, f] \quad \text{for } \varphi = 0,$$

and the axisymmetry of each of the two sums (see (3.19)) gives the equality everywhere in $\mathbb{R}^3$. Consequently we obtain

$$p^*[-v, f] = p^*[v, f],$$

as required.

As for (ii), we will show that

$$\left(\sum_{i,j=1}^{3} \partial_i u_j[v, f] \partial_j u_i[v, f] \right)(x_1, \rho) = \left(\sum_{i,j=1}^{3} \partial_i u_j[v, f] \partial_j u_i[v, f] \right)(-x_1, \rho), \qquad x_1 \in \mathbb{R}, \rho > 0,$$
(3.42)

where we skipped the φ in the variable (recall that this sum is independent of φ due to the axisymmetry (3.19)). In other words, the sum

$$\sum_{i,j=1}^{3} \partial_i u_j[v, f] \partial_j u_i[v, f]$$

is an even function (recall that in cylindrical coordinates $\rho = \sqrt{x_2^2 + x_3^2}$ takes the same value for x and $-x$) and so consequently $p^*[v, f]$ is even on $\mathbb{R}^3$ (by definition, see (3.33)). Then, in particular, $p[v, f]$ is even on $\mathbb{R}^2$, as required. Thus, it suffices to show (3.42).

To this end take $(-x_1, \rho, 0)$ as the variable in (3.41) to obtain the same expressions as in the case of $(x_1, \rho, 0)$, except for the diagonal expressions, which are now of the opposite sign. This, however, makes no change to the sum

$$\sum_{i,j=1}^{3} \partial_i u_j[v, f] \partial_j u_i[v, f],$$

that is,

$$\left(\sum_{i,j=1}^{3} \partial_i u_j[v, f] \partial_j u_i[v, f] \right)(x_1, \rho, 0) = \left(\sum_{i,j=1}^{3} \partial_i u_j[v, f] \partial_j u_i[v, f] \right)(-x_1, \rho, 0),$$

and thus (3.42) follows from the axisymmetry.   □

Finally, we point out that $u[v, f]$ amd $p[v, f]$ enjoy some useful properties regarding continuity with respect to f. The point is that given a sequence of v_k's and f_k's one can obtain convergence of

$$u[v_k, f_k] \cdot \Delta u[v_k, f_k] - u[v_k, f] \cdot \Delta u[v_k, f]$$

and

$$\nabla p[v_k, f_k] - \nabla p[v_k, f]$$

given convergence $f_k \to f$ (i.e. no convergence on v_k is required). The proof of such a result is easy but technical (involving some calculations in cylindrical coordinates) and since we will only use it (in (3.83), (4.36) and (3.82), (4.35), respectively) in a rather specific setting, we discuss it only in Sect. 3.6 at the end of the chapter.

3.2.4 A Structure on $U \Subset P$

The definitions in the previous section give rise to a way of defining a smooth, divergence-free velocity field u supported on $R(\overline{U})$, for $U \Subset P$. The following notion of a structure is a part of our simplified approach to the constructions.

Definition 3.4 A *structure on $U \Subset P$* is a triple (v, f, ϕ), where $v \in C_0^\infty(U; \mathbb{R}^2)$, $f \in C_0^\infty(P; [0, \infty))$, $\phi \in C_0^\infty(U; [0, 1])$ are such that supp $f = \overline{U}$,

$$\operatorname{supp} v \subset \{\phi = 1\}, \qquad \operatorname{div}(x_2 \, v(x_1, x_2)) = 0 \ in \ U \qquad and$$
$$f > |v| \ in \ U \ with \ Lf > 0 \ in \ U \setminus \{\phi = 1\}.$$

Note that (av, f, ϕ) is a structure for any $a \in (-1, 1)$ whenever (v, f, ϕ) is.

Furthermore, given (v, f, ϕ), a structure on U, the velocity field $u[v, f]$ is divergence free and is supported in $R(\overline{U})$. Moreover, in $R(\{\phi < 1\})$

$$u[v, f] \cdot \Delta u[v, f] \geq 0 \tag{3.43}$$

and

$$u[v, f] \cdot \nabla q = 0 \tag{3.44}$$

for any rotationally symmetric function $q \colon \mathbb{R}^3 \to \mathbb{R}$. This last property is particularly useful when taking $q := |u[v, f]|^2 + 2p[v, f]$ as in this way the left-hand side of (3.44) is of the same form as one of the terms in the Navier–Stokes inequality (3.5). In order to see (3.43), (3.44) first note that due to axisymmetry it is enough to verify that

$$u[v, f](x_1, x_2, 0) \cdot \Delta u[v, f](x_1, x_2, 0) \geq 0$$

and

$$u[v, f](x_1, x_2, 0) \cdot \nabla q(x_1, x_2, 0) = 0$$

for $(x_1, x_2) \in \{\phi < 1\}$ (recall (3.19)). Since $v = 0$ in $\{\phi < 1\}$ we have $u[v, f](x_1, x_2, 0) = (0, 0, f(x_1, x_2))$ (recall (3.24)), and so obtain the first of the above properties by writing

$$u[v, f](x_1, x_2, 0) \cdot \Delta u[v, f](x_1, x_2, 0) = f(x_1, x_2) Lf(x_1, x_2) \geq 0, \tag{3.45}$$

where we used Lemma 3.2 (ii). The second property follows in the same way by noting that $\partial_{x_3} q(x_1, x_2, 0) = 0$ (as a property of an axisymmetric function, which can be obtained in the same way as (3.31)).

Furthermore, note that given U, the L^∞ norm of derivatives of $u[v, f]$ can be bounded above by a constant depending only on $W^{1,\infty}$ norm of v and f, that is,

$$\|\nabla u[v, f]\|_{L^\infty} \le C \left(\|v\|_{W^{1,\infty}}, \|f\|_{W^{1,\infty}} \right), \qquad (3.46)$$

see (3.41). Note also that the constant depends on U only in terms of its distance from the x_1 axis.

3.2.5 A Recipe for a Structure

In the rest of the chapter, we will only consider functions v, f and sets $U \Subset P$ such that for some ϕ the triple (v, f, ϕ) is a structure on U. Moreover, we will only consider sets U in the shape of a rectangle or a "rectangular ring", that is, $V \setminus \overline{W}$, where V, W are open rectangles and $W \Subset V$. One can construct structures on such sets in a generic way, which we now describe.

First construct $v \in C_0^\infty(U; \mathbb{R}^2)$ satisfying div $(x_2 v(x_1, x_2)) = 0$ for all $(x_1, x_2) \in U$. For this it is enough to take a mollification of w and divide it by x_2, where $w : U \to \mathbb{R}^2$ is a compactly supported and weakly divergence-free vector field, that is $\int_P w \cdot \nabla \psi = 0$ for every $\psi \in C_0^\infty(P; \mathbb{R})$. Indeed, then the mollification of w is divergence free and thus div $(x_2 v(x_1, x_2)) = 0$. As for the construction of w take, for example,

$$w := (x_2 - 3, x_1) \chi_{1 < |x - (0,3)| < 2},$$

where χ denotes the indicator function, see Fig. 3.3. Note that w is weakly divergence free due to the fact that $w \cdot n$ vanishes on the boundary of the support of w, where n denotes the respective normal vector to the boundary. Alternatively, define w to be a "regionwise" constant velocity field

$$w := \begin{cases} (1, 0) & \text{in } R_1, \\ (0, 1) & \text{in } R_2, \\ (-1, 0) & \text{in } R_3, \\ (0, -1) & \text{in } R_4, \end{cases}$$

where R_1, R_2, R_3 and R_4 are arranged as in Fig. 3.3.

An integration by parts and the use of the crucial property of $w \cdot n$ being continuous across the boundary between each pair of neighbouring regions R_1, R_2, R_3, R_4, $P \setminus \bigcup_i R_i$ immediately shows that such a w is weakly divergence free. An advantage of such a definition of w (as compared to the previous one) is that it can be "stretched geometrically" in a sense that given $\varepsilon > 0$ one can modify w to obtain $w = (1, 0)$ in any given strict subset of P and $|w| < \varepsilon$ whenever v has a direction other than $(1, 0)$, see Fig. 3.4. We will later see an important sharpening of this observation (see Lemma 3.11).

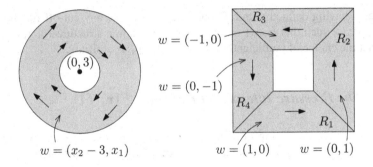

Fig. 3.3 Constructing compactly supported, weakly divergence-free vector field w

Fig. 3.4 Deforming the vector field w

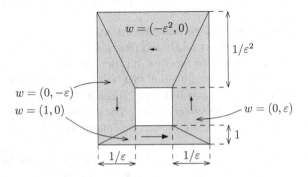

Second, let $\mu, \eta > 0$ be such that $\mu > \|v\|_\infty$ and supp $v \subset U_\eta$, where

$$U_\eta := \{x \in U : \text{dist}\,(x, \partial U) > \eta\} \tag{3.47}$$

denotes the η-*subset* of U, and let $f \in C_0^\infty(P; [0, \infty))$ be a certain cut-off function (in U) that has a particular behaviour near ∂U. Namely, let f be given by the following theorem.

Theorem 3.5 *Let $U \Subset P$ be an open set that is in the shape of a rectangle or $U = V \setminus \overline{W}$ for some open rectangles $V, W \in P$ with $W \Subset V$. Given $\eta > 0$ there exists $\delta \in (0, \eta)$ and $f \in C_0^\infty(P; [0, 1])$ such that*

$$\text{supp } f = \overline{U}, \quad f > 0 \text{ in } U \text{ with } f = 1 \text{ on } U_\eta$$

and

$$Lf > 0 \quad \text{in } U \setminus U_\delta.$$

The proof of the theorem is elementary in nature, but requires some technicalities, in particular, a generalised form of the Mean Value Theorem (see Lemma 3.13). We prove the theorem in Sect. 3.6 (see Lemma 3.15 for the case of a rectangle and Lemma 3.16 for the case of a rectangular ring).

Finally, having defined v and f, one can simply take any cut-off function $\phi \in C_0^\infty(U; [0, 1])$ such that $\phi = 1$ on U_δ. Thus, we obtain a structure (v, f, ϕ) on U. Note that the choice of (sufficiently large) $\mu = \|f\|_\infty$ is arbitrary.

3.2.6 The Pressure Interaction Function $F[v, f]$

As in the case of the notion of a structure (v, f, ϕ) on a set $U \Subset P$, we simplify Scheffer's approach by introducing the notion of a *pressure interaction function corresponding to* U,

$$F[v, f] := \nabla p[0, f] - \nabla p[v, f], \tag{3.48}$$

where ∇ denotes the two-dimensional gradient. Note that $F[v, f]$ depends on the structure (v, f, ϕ) on U, and thus a set $U \Subset P$ can possibly have more than one pressure interaction function. It is not clear whether $F[v, f]$ has any physical interpretation, but this is the tool that will form certain interactions between subsets of P (see the comments following Theorem 3.9), and we will see later that, in a sense, the strength of this interaction can be adjusted by manipulating the subsets and their corresponding structures (see the comments following (3.100) and the subsequent Sects. 3.4.2–3.4.5).

We now show that $F[v, f]$ enjoys a number of useful properties, which include estimates of its size at points near the x_1 axis.

Lemma 3.6 (*Properties of the pressure interaction function $F[v, f]$*) *Let (v, f, ϕ) be a structure on some $U \Subset P$ such that $v_1 \not\equiv 0$. Then the pressure interaction function $F := F[v, f]$ satisfies*

(i) $F \in C^\infty(\mathbb{R}^2; \mathbb{R}^2)$ *and*

$$\lim_{x_1 \to \pm\infty} x_1^4 F_1(x_1, 0) = \frac{\pm 9}{4\pi} \int_{\mathbb{R}^3} v_1^2 \left(y_1, \sqrt{y_2^2 + y_3^2} \right) dy =: \pm D.$$

(ii) F_1 *restricted to the x_1 axis attains a positive maximum, that is, there exists $B > 0$, $A \in \mathbb{R}$ such that*

$$B = F_1(A, 0) = \max_{x_1 \in \mathbb{R}} F_1(x_1, 0).$$

(iii) *There exists $C > 0$ such that*

$$|F(x)| \le C/|x|^4, \quad |\nabla F(x)| \le C/|x|^5 \quad \text{for } x \in \mathbb{R}^2.$$

(*Note $C \ge D$.*)

(iv) $F_2(x_1, 0) = 0$ *for $x_1 \in \mathbb{R}$.*

(v) *Let*

$$\kappa := 10^4 C/D. \tag{3.49}$$

There exists $N > 0$ such that for $n \geq N$

$$|x_1 - n| < \kappa, \ |x_2| < 1 \text{ implies } |F_1(x_1, x_2) - n^{-4}D| \leq 0.001 n^{-4} D.$$

Proof Claim (ii) follows from (i) and the assumption $v_1 \not\equiv 0$. As for (i), the smoothness of F follows directly from the fact that (v, f, ϕ) is a structure on U, and the limiting property as $x_1 \to \pm\infty$ follows by using (3.21), from which we obtain

$$\lim_{x_1 \to \infty} x_1^4 F_1(x_1, 0) = \frac{3}{4\pi} \int_{\mathbb{R}^3} \left(|u[0, f]|^2 - 3(u_1[0, f])^2 - |u[v, f]|^2 + 3(u_1[v, f])^2 \right) dy$$

$$= \frac{9}{4\pi} \int_{\mathbb{R}^3} v_1(R^{-1}(y)) dy,$$

where we also used the facts $|u[v, f](y)| = f(R^{-1}(y))$, $u_1[v, f](y) = v_1(R^{-1}(y))$ (see (3.26)). The case of the limit $x_1 \to -\infty$ is similar.

Claim (iii) follows from the decay properties of the pressure function, see (3.20). Claim (iv) follows directly from (3.37).

As for (v), suppose that $|x_1 - n| < \kappa$. Then for sufficiently large n (and so also x_1)

$$|n^4 - x_1^4| = |n^2 + x_1^2| \, |n + x_1| \, |n - x_1| \leq \tilde{C} |x_1|^3$$

for some $\tilde{C} > 0$ (depending on κ). Thus

$$|n^4 F_1(x_1, 0) - D| \leq |n^4 - x_1^4| \, |F_1(x_1, 0)| + |x_1^4 F_1(x_1, 0) - D| \leq \tilde{C} C |x_1|^{-1} + |x_1^4 F_1(x_1, 0) - D|.$$

Since taking n large makes x_1 large as well, we see from (i) that for sufficiently large n $|n^4 F_1(x_1, 0) - D| \leq 0.0005D$, that is,

$$|F_1(x_1, 0) - n^{-4} D| \leq 0.0005 n^{-4} D. \tag{3.50}$$

Moreover, the Mean Value Theorem gives for $|x_2| < 1$ and sufficiently large n

$$|F_1(x_1, x_2) - F_1(x_1, 0)| \leq |x_2| \, |\nabla F_1(x_1, \xi)| \leq C |x_1|^{-5} \leq 0.0005 n^{-4} D,$$

where $\xi \in (0, 1)$. The claim follows from this and (3.50). $\qquad\square$

3.3 The Setting

In this section, we define constants $T > 0$, $\nu_0 > 0$, $\tau \in (0, 1)$, $z \in \mathbb{R}^3$, the set G and the vector field u which were required in the sketch proof in Sect. 3.1. The definition is based on a certain geometric setting which we formalise here in the notion of the *geometric arrangement*.

By the *geometric arrangement* we mean a pair of open sets $U_1, U_2 \Subset P$ together with the corresponding structures $(v_1, f_1, \phi_1), (v_2, f_2, \phi_2)$ (recall Definition 3.4) such that $\overline{U_1} \cap \overline{U_2} = \emptyset$ and, for some $T > 0, \tau \in (0, 1), z \in \mathbb{R}^3$,

$$f_2^2 + Tv_2 \cdot F[v_1, f_1] > |v_2|^2 \quad \text{in } U_2, \tag{3.51}$$

$$f_2^2(y) + Tv_2(y) \cdot F[v_1, f_1](y) > \tau^{-2} \left(f_1(R^{-1}x) + f_2(R^{-1}x) \right)^2 \tag{3.52}$$

for all $x \in G := R\left(\overline{U_1} \cup \overline{U_2}\right)$, where

$$y = R^{-1}(\Gamma(x)) \tag{3.53}$$

(recall $\Gamma(x) = \tau x + z$).

Note that (3.52) gives, in particular, that Γ maps G into itself: we obtain $R^{-1}(\Gamma(G)) \subset G$ and so (taking the rotation R of both sides)

$$\Gamma(G) \subset RR^{-1}(G) \subset R(G) = G. \tag{3.54}$$

Before defining the remaining constant ν_0 and vector field u, we comment on the notion of the geometric arrangement in an informal way.

Recall from Sect. 3.1 that we aim to find a vector field u, which is defined on the time interval $[0, T]$, that satisfies the NSI (3.5) as well as admits the gain in magnitude (3.6). We want to obtain the gain via the term $u \cdot \nabla p$, which we now discuss. We will construct u in a way that, at time $t = 0$

$$u(0) \approx u[v_1, f_1] + u[v_2, f_2],$$

and at time $t = T$

$$u(T) \approx u[v_1, f_1] + u\left[v_2, \sqrt{f_2^2 + Tv_2 \cdot F[v_1, f_1]} \right]. \tag{3.55}$$

In other words, u consists of two disjointly supported (in space) vector fields. The first of them will be supported in $R(\overline{U_1})$ and its absolute value (that is f_1) will remain (approximately) constant through the time interval $[0, T]$. The second of them will be supported in $R(\overline{U_2})$ and its absolute value will change in time from f_2 to (approximately) $\sqrt{f_2^2 + Tv_2 \cdot F[v_1, f_1]}$.

At this point, it is clear that the requirement (3.51) is necessary for the right-hand side of (3.55) to be well defined (recall (3.23)). Furthermore, in light of the property $|u[v, f]| = f$ (valid for any (admissible) v, f, recall (3.25)) we see that the requirement (3.52) means simply that

$$|u(\Gamma(x), T)|^2 \gtrsim \tau^{-2} |u(x, 0)|^2.$$

By writing "approximately" (or $\approx$, $\gtrsim$) we mean "very close in the $L^\infty(\mathbb{R}^3)$ norm". Such an approximate sense will be made rigorous below by using continuity arguments as well as the facts that the inequalities in (3.51) and (3.52) are sharp (">") and the supports of the functions appearing on their right-hand sides are compact.

It remains to ask why the term "$T v_2 \cdot F[v_1, f_1]$" is chosen to achieve the gain in magnitude.

A rough answer to this question is because (1) the pressure interaction function has a certain property that allows us to magnify it and because (2) this is one of the very few degrees of freedom allowed by the Navier–Stokes inequality. We have already observed (1) in Lemma 3.6 (particularly part (ii)), and we will see the full power of it in the construction of the geometric arrangement in Sect. 3.4. As for (2), recall the NSI (3.5),

$$\partial_t |u|^2 \le -u \cdot \nabla \left(|u|^2 - 2p\right) + 2\nu\, u \cdot \Delta u.$$

We illustrate the reason for the term "$T v_2 \cdot F[v_1, f_1]$" by the following thought experiment. Suppose that

$$u = u[v_1, f_1] + u[v_2, f_2] \tag{3.56}$$

and take a close look at the terms appearing on the right-hand side of the NSI above, where we ignore, for a moment, the time dependence. First of all, the pressure function p is given by $p^*[v_1, f_1] + p^*[v_2, f_2]$ (recall Lemma 3.3 (iii)). Thus, since both u and p are axisymmetric, so are all the terms on the right-hand side of the NSI (recall (3.19)). Thus, it is sufficient to look only at points x of the form $(x_1, x_2, 0)$, $(x_1, x_2) \in \mathbb{R}^2$. At such points, the right-hand side of the NSI takes the form

$$- \{(v_1 + v_2) \cdot \nabla\} (f_1^2 + f_2^2 + 2p[v_1, f_1] + 2p[v_2, f_2]) + 2\nu u(\cdot, 0) \cdot \Delta u(\cdot, 0), \tag{3.57}$$

where $v_1 = (v_{11}, v_{12})$, $v_2 = (v_{21}, v_{22})$ and $\nabla = (\partial_1, \partial_2)$ now denotes the two-dimensional gradient; recall also that $p[v_i, f_i] = p^*[v_i, f_i](\cdot, 0)$ (see (3.34)). Observe that the ∂_3 derivative does not appear since both u and p are axisymmetric (and so ∂_3 is a derivative along a level set, recall (3.31) and (3.36)).

The last term in (3.57) will not play any significant role in our analysis; we will treat it as an error term. In fact, we already know how to deal with this term at points (x_1, x_2) such that $(\phi_1 + \phi_2)(x_1, x_2) < 1$ (recall that ϕ_1, ϕ_2 play the role of a cut-off function in the structures (v_1, f_1, ϕ_1), (v_2, f_2, ϕ_2), respectively; see Definition 3.4). Indeed, at such points $v_1 = v_2 = 0$, and so (3.57) becomes

$$2\nu(f_1\, Lf_1 + f_2\, Lf_2) \ge 0$$

(the inequality being a consequence of (3.45)). This non-negativity will turn out sufficient for the NSI (see (3.81) below for details), while at points (x_1, x_2) such that $(\phi_1 + \phi_2)(x_1, x_2) = 1$ we will use continuity arguments to take ν sufficiently small (see (3.71) and (3.84) below for details).

As for the first term in (3.57), we will be interested in interactions between $u[v_1, f_1]$ and $u[v_2, f_2]$ (this is the reason why the geometric arrangement consists of two sets U_1, U_2 and their corresponding structures) and so from the terms in (3.57) we are concerned with the mixed terms of the form

(something supported in U_i) (a function "based on" U_j and its structure)

for $i, j = 1, 2, i \neq j$, namely, with the terms

$$- v_i \cdot \nabla f_j^2 \quad \text{and} \quad - 2v_i \cdot \nabla p[v_j, f_j], \tag{3.58}$$

$i, j = 1, 2, i \neq j$. Note that the first of such terms vanishes since v_i and f_j have disjoint supports. As for the second one, we will be manipulating only the terms with the "∂_1" derivative since we are only able to control this derivative of the pressure function (which comes, fundamentally, from the property (3.21) and from our choice of picking Ox_1 as the axis of symmetry; we have already explored this (to some extent) in Lemma 3.6). In fact, we aim to construct the geometric arrangement in such a way that

$$- v_{21}\partial_1(p[v_1, f_1] - p[0, f_1]) = v_{21} F_1[v_1, f_1] \quad \text{is large} \tag{3.59}$$

in a certain region of U_2 that is close to the Ox_1 axis (see Sect. 3.4.1 for a wider discussion of this issue). In other words, we will try to, in a sense, magnify the influence of U_1 (and its structure) onto U_2 (and its structure).

We now discuss the issue of time dependence, which will explain the appearance of the term "$p[0, f_1]$" in (3.59) above as well as the mechanism responsible to "picking" (3.59) from all possible choices at the second term in (3.58) (as i, j varies). This will lead us to the term "$Tv_2 \cdot F[v_1, f_1]$" in the geometric arrangement. In fact, instead of the naive candidate (3.56), we will actually consider a time-dependent vector field of the form

$$u(t) = u[v_{1,t}, f_{1,t}] + u[v_{2,t}, f_{2,t}],$$

where $v_{i,t}$, $f_{i,t}$ are certain time-dependent extensions of v_i, f_i, respectively (see (3.75) and (3.74) for the exact formula), which are chosen so that

$$|u(\cdot, 0, t)|^2 = f_{1,t}^2 + f_{2,t}^2 = f_1^2 + f_2^2 + (\text{something small, negative and linear in } t)$$

$$- \int_0^t (v_{1,s} + v_{2,s}) \cdot \nabla(f_{1,s}^2 + f_{2,s}^2 + 2p[v_{1,s}, f_{1,s}] + 2p[v_{2,s}, f_{2,s}])ds, \tag{3.60}$$

(We write "$(\cdot, 0, t)$" to articulate that we restrict ourselves to points (x, t) of the form $(x_1, x_2, 0, t)$.) Note that by taking ∂_t we obtain

$$\partial_t |u(\cdot, 0, t)|^2 = - \text{(something small)}$$
$$- (v_{1,t} + v_{2,t}) \cdot \nabla(f_{1,t}^2 + f_{2,t}^2 + 2p[v_{1,t}, f_{1,t}] + 2p[v_{2,t}, f_{2,t}]),$$
$$= - \text{(something small)} - u(\cdot, 0, t) \cdot \nabla(|u(\cdot, 0, t)|^2 + 2p(\cdot, 0, t)).$$

Here, the small term will be used in the continuity argument to absorb the Laplacian term, $\nu u \cdot \Delta u$ (compare with (3.57)), see (3.81) and (3.84) for details. In other words, the time-dependent extensions $v_{i,t}$, $f_{i,t}$ ($i = 1, 2$) will be chosen such that, by construction, we will obtain the NSI.

In particular, we will choose

$$v_{i,t} = a_i(t)v_i, \qquad i = 1, 2,$$

where $a_1, a_2 \in C^\infty(\mathbb{R}; [-1, 1])$ are certain *oscillatory processes*, which are discussed in detail in Sect. 3.3.3 below. The oscillatory process will have two remarkable features. The first is that

$$\int_0^t a_i(s)v_i \cdot \nabla f_{i,s} ds \approx 0, \qquad \text{uniformly in } i = 1, 2, t \in [0, T],$$

and it will be a simple consequence of high oscillations of a_1, a_2. The second remarkable feature is that they enable us to pick from all the terms

$$\int_0^t a_i(s)v_i \cdot \nabla p[a_j(s)v_j, f_{j,s}] ds, \qquad i, j \in \{1, 2\}$$

any of the terms

$$\int_0^t v_i \cdot \nabla p[v_j, f_{j,s}],$$

provided we subtract $p[0, f_{j,s}]$. To be more precise for any choice of indices $i_0, j_0 \in \{1, 2\}$ there exist oscillatory processes $a_1, a_2 \in C^\infty(\mathbb{R}; [-1, 1])$ such that

$$\sum_{i,j=1}^2 \int_0^t a_i(s)v_i \cdot \nabla p[a_j(s)v_j, f_{j,s}] ds \approx \int_0^t v_{i_0} \cdot \nabla(p[v_{j_0}, f_{j_0,s}] - p[0, f_{j_0,s}]) ds \quad \text{for all } t \in [0, T].$$

Therefore, choosing $(i_0, j_0) := (2, 1)$ (since we are interested in the influence of U_1 onto U_2) we obtain that the integral on the right-hand side of (3.60) is approximately

$$\int_0^t v_2 \cdot \nabla(p[v_1, f_{1,s}] - p[0, f_{1,s}]) ds = - \int_0^t v_2 \cdot F[v_1, f_{1,s}] ds \quad \text{for all } t \in [0, T].$$

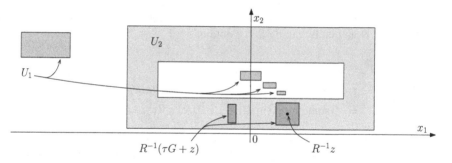

Fig. 3.5 A sketch of the geometric arrangement, see Sect. 3.4 for details. Note that the inclusion $R^{-1}(\tau G + z) \subset U_2 \subset R^{-1}(G)$, which is illustrated on this sketch, implies that $\Gamma(G) \subset G$ (recall (3.54)). Proportions are not conserved on this sketch

On the other hand, we will make a choice of $f_{1,s}$, that is, roughly speaking, very slowly depending on s, so that the last integral is approximately

$$t\, v_2 \cdot F[v_1, f_1].$$

In other words, we will choose the oscillatory processes a_1, a_2 and the time-dependent extensions of f_1, f_2 such that, except for the expression of $|u(t)|$ given (approximately) by (3.60), we will obtain, at the same time, another one:

$$|u(\cdot, 0, t)|^2 \approx f_1^2 + f_2^2 + t\, v_2 \cdot F[v_1, f_1] \quad t \in [0, T]. \tag{3.61}$$

This explains (by taking $t = T$) the appearance of the term $T v_2 \cdot F[v_1, f_1]$ in the geometric arrangement.

To sum up the above heuristic discussion, based on any disjoint sets U_1, U_2 and their corresponding structures (v_1, f_1, ϕ_1), (v_2, f_2, ϕ_2), we can find a way of prescribing the time dependence (on any time interval) such that the NSI is satisfied (by prescribing behaviour in time, in particular, by the oscillatory processes) and that $|u(t)|$ is approximately as in (3.61), which in turn we are able to magnify (at least in some region of the support) by arranging U_1, U_2 (and the corresponding structures) and defining $T > 0$ appropriately; namely, by constructing the geometric arrangement.

The construction of the geometric arrangement (which is sketched in Fig. 3.5) is a nontrivial matter and it is in fact the heart of the proof of Theorem 3.1. We present it in Sect. 3.4.

In the remainder of this section, we assume that the geometric arrangement is given and we apply the strategy outlined in the heuristic discussion above, but in a rigorous way. Namely, we obtain ν_0 and u (the remaining constants $T > 0, \tau \in (0, 1)$, $z \in \mathbb{R}^3$ and the compact set $G \subset \mathbb{R}^3$, which were required in the sketch argument in Sect. 4.2, are given by the geometric arrangement).

We note that, except for the need of rigorous presentation (in the remainder of this section as well as in Sect. 3.4, where we construct the geometric arrangement), it is also rather pleasing to observe all components of the construction fit together.

Furthermore, we will not be using the notation $f_{i,t}$ (to denote the time extension of f_i, $i = 1, 2$), but rather $h_{i,t}$ (the time extension of f_i) and $q_{i,t}^k$ (an approximation of $h_{i,t}$, where k is large).

Let $\theta > 0$ be sufficiently small such that

$$f_2^2(y) + Tv_2(y) \cdot F[v_1, f_1](y) > \tau^{-2}\left(f_1(R^{-1}x) + f_2(R^{-1}x)\right)^2 + 2\theta \quad (3.62)$$

for $x \in G$. (Recall $y = R^{-1}(\Gamma x)$.) Such a choice is possible by continuity since the inequality in (3.52) is strict and G is compact.

Let $h: P \times [0, T] \to [0, \infty)$ be defined by

$$h_t = h_{1,t} + h_{2,t} \quad (3.63)$$

(recall we use the convention $h_t(\cdot) \equiv h(\cdot, t)$), where

$$h_{1,t}^2 := f_1^2 - 2t\delta\phi_1, \quad (3.64)$$

$$h_{2,t}^2 := f_2^2 - 2t\delta\phi_2 + \int_0^t v_2 \cdot F[v_1, h_{1,s}]\,ds. \quad (3.65)$$

Thus, $h_{i,t}$ is a time-dependent modification of f_i, $i = 1, 2$, such that $h_{i,t} = f_i$ outside $\operatorname{supp}\phi_i$ (recall $\operatorname{supp}v_2 \subset \operatorname{supp}\phi_2$, see Definition 3.4). Here $\delta > 0$ is a fixed, small number given by the following lemma.

Lemma 3.7 (properties of functions $h_{1,t}, h_{2,t}$) *There exists $\delta > 0$ (sufficiently small) such that $h_1, h_2 \in C^\infty(P \times (-\delta, T + \delta); [0, \infty))$,*

$$(v_i, h_{i,t}, \phi_i) \text{ is a structure on } U_i \quad \text{for } t \in (-\delta, T + \delta), i = 1, 2, \quad (3.66)$$

and

$$h_{2,T}^2(y) > \tau^{-2}\left(f_1(R^{-1}x) + f_2(R^{-1}x)\right)^2 + \theta \quad \text{for } x \in R\left(\overline{U_1} \cup \overline{U_2}\right). \quad (3.67)$$

(Recall $y = R^{-1}(\Gamma(x))$.)

Proof For $h_{1,t}$ note that since $f_1 > 0$ in U_1 we can take $\delta \in (0, 1)$ such that

$$\delta < \min_{\operatorname{supp}\phi_1}\left|f_1^2 - |v_1|^2\right|/2(T + 2)$$

to obtain

$$h_{1,t} > |v_1| \quad \text{in supp}\, \phi_1 \text{ for } t \in [-1, T+1].$$

Thus, since $h_{1,t} = f_1$ outside supp ϕ_1,

$$h_{1,t} > |v_1| \geq 0 \quad \text{in } U_1 \text{ for } t \in (-\delta, T+\delta).$$

Hence, since both f_1 and ϕ_1 are smooth on P we immediately obtain the required smoothness of h_1 and that $(v_1, h_{1,t}, \phi_1)$ is a structure on U_1 for all $t \in (-\delta, T+\delta)$.

As for $h_{2,t}$, suppose for the moment that $\delta = 0$. Then $h_{1,t} = f_1$ and so

$$h_{2,t}^2 = f_2^2 + t\, v_2 \cdot F[v_1, f_1]. \tag{3.68}$$

This means that

$$h_{2,0}^2 = f_2^2 \quad \text{and} \quad h_{2,T}^2 = f_2^2 + T v_2 \cdot F[v_1, f_1] \quad \text{if } \delta = 0. \tag{3.69}$$

Using the fact that (v_2, f_2, ϕ_2) is a structure on U_2 and (3.51), we see that both of the above functions are greater than $|v_2|^2$ in U_2. In particular, they are greater than $|v_2|^2$ on the compact set supp ϕ_2. Since $h_{2,t}$ in (3.68) depends linearly on t, we thus obtain

$$h_{2,t} > |v_2| \quad \text{in supp}\, \phi_2, \text{ for } t \in [0, T] \quad \text{if } \delta = 0.$$

Therefore, since h_2 depends continuously on δ we obtain

$$h_{2,t} > |v_2| \quad \text{in supp}\, \phi_2, \text{ for } t \in [0, T] \quad \text{if } \delta > 0 \text{ is sufficiently small.}$$

Thus, by continuity in time, this property holds also for t belonging to an open interval containing $[0, T]$. Taking δ smaller we can take this open interval to be $(-\delta, T+\delta)$. Thus, recalling that $h_{2,t} = f_2$ outside supp ϕ_2 we obtain

$$h_{2,t} > |v_2| \geq 0 \quad \text{in } U_2, \text{ for } t \in (-\delta, T+\delta) \quad \text{if } \delta > 0 \text{ is sufficiently small.}$$

As in the case of $h_{1,t}$, this immediately gives the required regularity of h_2 and that $(v_2, h_{2,t}, \phi_2)$ is a structure on U_2.

As for (3.67) note that (3.69) gives, in particular,

$$h_{2,T}^2 = f_2^2 + T v_2 \cdot F[v_1, f_1] \quad \text{in supp}\, \phi_2 \quad \text{if } \delta = 0,$$

and so for sufficiently small $\delta > 0$

$$h_{2,T}^2 \geq f_2^2 + T v_2 \cdot F[v_1, f_1] - \theta \quad \text{in supp}\, \phi_2.$$

Since $h_{2,T} = f_2$ and $v_2 = 0$ outside supp ϕ_2, we trivially obtain the above inequality outside supp ϕ_2, and so (3.67) follows from this and (3.62). $\qquad\square$

We have now fixed $\delta > 0$. Note that (3.66) gives, in particular, that

$$(av_i, h_{i,t}, \phi_i) \text{ is a structure on } U_i \qquad \text{for } t \in (-\delta, T + \delta), i = 1, 2, \qquad (3.70)$$

for any $a \in [-1, 1]$ $(t \in (-\delta, T + \delta), i = 1, 2)$.

At this point, we fix $\nu_0 \in (0, 1)0$ sufficiently small such that

$$\nu_0 \left| u[av_i, h_{i,t}] \cdot \Delta u[av_i, h_{i,t}] \right| < \delta/8 \quad \text{in } \mathbb{R}^3 \qquad (3.71)$$

for $a \in [-1, 1]$, $i = 1, 2$, $t \in [0, T]$.

Having constructed the time-dependent functions $h_{1,t}, h_{2,t}$ and having fixed ν_0, we now construct u.

Proposition 3.8 *There exist $\eta \in (0, \delta)$ and $u \in C^\infty(\mathbb{R}^3 \times (-\eta, T + \eta); \mathbb{R}^3)$ such that*

(i) $\operatorname{supp} u(t) = G$ and $\operatorname{div} u(t) = 0$ for $t \in (-\eta, T + \eta)$,
(ii) $|u(x, 0)| = h_0(R^{-1}x)$ and $\left| |u(x, t)|^2 - h_t(R^{-1}x)^2 \right| < \theta$ for all $x \in \mathbb{R}^3$, $t \in [0, T]$,
(iii) the Navier–Stokes inequality

$$\partial_t |u|^2 \leq -u \cdot \nabla \left(|u|^2 + 2p \right) + 2\nu\, u \cdot \Delta u$$

holds in $\mathbb{R}^3 \times [0, T]$ for all $\nu \in [0, \nu_0]$ where p is the pressure function corresponding to u.

Note that u given by the proposition satisfies all the properties required in Sect. 3.1. Among those only (3.6) is nontrivial; this follows from (ii) and (3.67) by writing

$$\begin{aligned}
|u(\Gamma(x), T)|^2 &\geq h_T(R^{-1}(\Gamma(x))^2 - \theta \\
&\geq h_{2,T}(R^{-1}(\Gamma(x))^2 - \theta \\
&> \tau^{-2} \left(f_1(R^{-1}x) + f_2(R^{-1}x) \right)^2 \\
&= \tau^{-2} h_0(R^{-1}(x))^2 \\
&= \tau^{-2} |u(x, 0)|^2
\end{aligned}$$

for $x \in G$ (the case $x \notin G$ is trivial). The rest of the properties follow directly from (i), (iii). It remains to prove Proposition 3.8. The proof is separated into three steps, which we present in Sects. 3.3.1–3.3.3.

3.3.1 The Construction of u

We will find u (a solution of Proposition 3.8) that is axisymmetric (see (3.18)). For such a vector field (ii) is equivalent to

$$|u(x, 0, 0)| = h_0(x), \quad \text{and} \quad \left| |u(x, 0, t)|^2 - h_t(x)^2 \right| < \theta \quad \text{for } x \in P, t \in [0, T]$$

$$(3.72)$$

and (iii) is equivalent to

$$\partial_t |u(x, 0, t)|^2 \leq -u(x, 0, t) \cdot \nabla \left(|u(x, 0, t)|^2 + 2p(x, 0, t) \right) + 2\nu\, u(x, 0, t) \cdot \Delta u(x, 0, t)$$

$$(3.73)$$

being satisfied for all $x \in P$, $t \in [0, T]$, $\nu \in [0, \nu_0]$.

We will consider functions q_1^k, q_2^k defined by

$$\left(q_{i,t}^k \right)^2 := f_i^2 - 2t\delta\phi_i - \int_0^t a_i^k(s) v_i \cdot \left(\nabla h_{i,s}^2 + 2\nabla p[a_1^k(s)v_1, h_{1,s}] + 2\nabla p[a_2^k(s)v_2, h_{2,s}] \right) ds,$$

$$(3.74)$$

$i = 1, 2$, $k \in \mathbb{N}$, for some functions $a_1^k, a_2^k \in C^\infty(\mathbb{R}; [-1, 1])$ (which we shall call *oscillatory processes* and which we discuss below). Recall that we use the convention $q_{i,t}^k(\cdot) \equiv q_i^k(\cdot, t)$ (see (3.15)). We will show that, given a particular choice of the oscillatory processes a_1^k, a_2^k, the vector field

$$u(x, t) := u[a_1^k(t)v_1, q_{1,t}^k](x) + u[a_2^k(t)v_2, q_{2,t}^k](x), \quad (3.75)$$

is a solution of Proposition 3.8 for sufficiently large k. Note that such u is axisymmetric (recall Sect. 3.2.3). Before proceeding to the proof, we comment on this strategy in an informal way.

Forget, for the moment, about the functions q_1^k, q_2^k, and let us try to attack Proposition 3.8 directly. We observe that part (ii) and the facts that $h_{1,t}, h_{2,t}$ have disjoint supports $\overline{U_1}, \overline{U_2}$ (respectively) and that $(v_1, h_{1,t}, \phi_1), (v_2, h_{2,t}, \phi_2)$ are structures on U_1, U_2 (respectively) suggest looking at the velocity field of the form

$$\widetilde{u}(x, t) := u[v_1, h_{1,t}](x) + u[v_2, h_{2,t}](x). \quad (3.76)$$

In other words, we have

$$|\widetilde{u}(x, 0, t)|^2 = h_{1,t}^2(x) + h_{2,t}^2(x) \quad x \in P, t \in [0, T],$$

so that claim (ii) is satisfied in an exact sense (rather than in an approximate sense with error θ). This might look promising, but, recalling the definition of h_1, h_2 (see (3.64), (3.65)), we see that

$$\partial_t |\widetilde{u}(x, 0, t)|^2 = -2\delta(\phi_1 + \phi_2)(x) + v_2(x) \cdot F[v_1, h_{1,t}](x),$$

and at this point it is not clear how to relate the right-hand side to the terms

$$-u \cdot \nabla \left(|u|^2 + 2p \right) + \nu u \cdot \Delta u,$$

which are required by the Navier–Stokes inequality (3.73) (that is by (iii)). Thus, the velocity field $\widetilde{u}$ seems unlikely to be a solution of Proposition 3.8. In order to proceed one needs to make use of two degrees of freedom available in the construction of $\widetilde{u}$. The first of them is the fact that claim (ii) of Proposition 3.8 only requires $|u(x,t)|$ to "keep close" to $h_t(R^{-1}x)$ as t varies between 0 and T (rather than to be equal to it), which we have already pointed out above. The second one is that $|\widetilde{u}(x,0,t)|$ is expressed only in terms of $h_{1,t}, h_{2,t}$. Thus, a velocity field of the form

$$\overline{u}(x,t) := u[a_1(t)v_2, h_{1,t}](x) + u[a_2(t)v_2, h_{2,t}](x)$$

has the same absolute value $|\overline{u}|$ as $|\widetilde{u}|$ for any choice of $a_1, a_2 : \mathbb{R} \to [-1, 1]$. Recall also that since $|a_1|, |a_2| \leq 1$,

$$\left(a_i(t)v_i, h_{i,t}, \phi_i\right) \quad \text{is a structure on } U_i, i = 1, 2,$$

(recall (3.70)) and so $\overline{u}$ is well defined. By introducing the functions q_1^k, q_2^k (in (3.74)) we make use of these two degrees of freedom.

We now proceed to a discussion of some elementary properties of these functions, and we show in Sect. 3.3.2 that considering them is a good idea, namely, that (3.75) is a solution of the proposition for sufficiently large k.

First note that, as in the case of $h_{i,t}$, $q_{i,t}^k$ differs from f_i only on the compact set supp ϕ_i, $i = 1, 2$. Second,

$$\partial_t \left(q_{i,t}^k\right)^2 = -2\delta\phi_i - a_i^k(t)v_i \cdot \left(\nabla h_{i,t}^2 + 2\nabla p[a_1^k(t)v_1, h_{1,t}] + 2\nabla p[a_2^k(t)v_2, h_{2,t}]\right). \tag{3.77}$$

(Compare the terms appearing on the right-hand side to (3.58).)

Finally, we will show in Sect. 3.3.3 that, given a particular choice of the oscillatory processes $a_1^k, a_2^k \in C^\infty(\mathbb{R}, [-1, 1])$ (which are a part of the definition of q_1^k, q_2^k, recall (3.74)),

$$\begin{cases} q_{i,t}^k \to h_{i,t} \\ \text{and} \\ D^l q_{i,t}^k \to D^l h_{i,t} \end{cases} \quad \text{uniformly in } P \times [0, T], i = 1, 2, \text{ for each } l \geq 1$$

$$\tag{3.78}$$

as $k \to \infty$. Recalling properties of $h_{1,t}, h_{2,t}$ (see Lemma 3.7), we see that this convergence gives, in particular, that for sufficiently large k

$$q_{i,t}^k > |v_i| \quad \text{in supp } \phi_i \quad \text{for } t \in [0, T], i = 1, 2,$$

and so by continuity (as in the proof of Lemma 3.7)

$$q_{i,t}^k > |v_i| \geq 0 \quad \text{in } U_i \quad \text{for } t \in (-\delta_k, T + \delta_k), \tag{3.79}$$

for some $\delta_k \in (0, \delta)$. Thus, for sufficiently large k

$$(v_i, q_{i,t}^k, \phi_i) \text{ is a structure on } U_i \quad \text{for } t \in (-\delta_k, T + \delta_k), i = 1, 2,$$

and thus, since $|a_1^k|, |a_2^k| \in [-1, 1]$,

$$(a_i^k(t)v_i, q_{i,t}^k, \phi_i) \text{ is a structure on } U_i \qquad \text{for } t \in (-\delta_k, T + \delta_k), i = 1, 2. \quad (3.80)$$

Moreover, (3.79) and the fact that all terms on the right-hand side of (3.74) are smooth (recall (3.35) for the smoothness of the pressure) give

$$q_i^k \in C^\infty(P \times (-\delta_k, T + \delta_k); [0, \infty)), \qquad i = 1, 2.$$

3.3.2 The Proof of the Claims of the Proposition

Using the above properties of the functions q_1^k, q_2^k, we now show that for k sufficiently large the vector field u given by (3.75),

$$u(x, t) := u[a_1^k(t)v_1, q_{1,t}^k](x) + u[a_2^k(t)v_2, q_{2,t}^k](x),$$

with $\eta := \delta_k$ satisfies the claims of Proposition 3.8.

Claim (i) and the smoothness of u on $\mathbb{R}^3 \times (-\eta, T + \eta)$ follow directly from (3.80), the smoothness of the oscillatory processes a_1^k, a_2^k on $\mathbb{R}$ (which we are about to construct in the next section) and from the smoothness of q_1^k, q_2^k stated above.

Claim (ii) is equivalent to (3.72) (due to axisymmetry of u), and thus its first part follows by writing

$$|u(x, 0, 0)| = q_{1,0}^k(x) + q_{2,0}^k(x) = f_1(x) + f_2(x) = h_0(x).$$

The second part follows directly from the convergence (3.78) by taking k sufficiently large such that

$$\left| (q_{1,t}^k + q_{2,t}^k)^2 - h_t^2 \right| < \theta \qquad \text{in } P, \quad \text{for } t \in [0, T].$$

For such k we obtain

$$\left| |u(x, 0, t)|^2 - h_t(x)^2 \right| = \left| (q_{1,t}^k(x) + q_{2,t}^k(x))^2 - h_t(x)^2 \right| < \theta,$$

as required.

As for claim (iii), first recall that $p(t)$, the pressure function corresponding to $u(t)$, is (due to (3.33)) given by

$$p(t) = p^*[a_1^k(t)v_1, q_{1,t}^k] + p^*[a_2^k(t)v_2, q_{2,t}^k],$$

and so, in particular,

$$p(x, 0, t) = p[a_1^k(t)v_1, q_{1,t}^k](x) + p[a_2^k(t)v_2, q_{2,t}^k](x).$$

Recalling that claim (iii) is equivalent to (3.73), that is, the Navier–Stokes inequality restricted to P,

$$\partial_t |u(x, 0, t)|^2 \leq -u(x, 0, t) \cdot \nabla \left(|u(x, 0, t)|^2 + 2p(x, 0, t) \right) + 2\nu \, u(x, 0, t) \cdot \Delta u(x, 0, t),$$

where $\nu \in [0, \nu_0]$ (recall (3.71) for the choice of ν_0), we fix $x \in P$, $t \in [0, T]$ and we consider two cases.

Case 1. $\phi_1(x) + \phi_2(x) < 1$.

For such x, we have $v_1(x) = v_2(x) = 0$ (from the elementary properties of structures, recall Definition 3.4) and the Navier–Stokes inequality follows trivially for all k by writing

$$\begin{aligned}
\partial_t |u(x, 0, t)|^2 &= \partial_t q^k_{1,t}(x)^2 + \partial_t q^k_{2,t}(x)^2 \\
&= -2\delta(\phi_1(x) + \phi_2(x)) \\
&\leq 0 \\
&\leq -u(x, 0, t) \cdot \nabla \left(|u(x, 0, t)|^2 + 2p(x, 0, t) \right) + 2\nu \, u(x, 0, t) \cdot \Delta u(x, 0, t),
\end{aligned}$$
(3.81)

where we used (3.43) and (3.44) in the last step.

Case 2. $\phi_1(x) + \phi_2(x) = 1$.

In this case, we need to use the convergence (3.78) to take k sufficiently large such that

$$|v_i| \left(\left| \nabla(q^k_{i,t})^2 - \nabla h^2_{i,t} \right| + 2 \sum_{j=1,2} \left| \nabla p[a^k_j(t)v_j, q^k_{j,t}] - \nabla p[a^k_j(t)v_j, h_{j,t}] \right| \right) \leq \delta/2$$
(3.82)

in P (see Lemma 3.17 for a verification that (3.78) is sufficient for the convergence of the pressure functions) and

$$\begin{aligned}
\nu_0 &\left| u[a^k_i(t)v_i, q^k_{i,t}] \cdot \Delta u[a^k_i(t)v_i, q^k_{i,t}] \right| \\
&\leq \nu_0 \left| u[a^k_i(t)v_i, h_{i,t}] \cdot \Delta u[a^k_i(t)v_i, h_{i,t}] \right| + \delta/8 \leq \delta/4
\end{aligned}$$
(3.83)

in $\mathbb{R}^3$, for $t \in [0, T]$, $i = 1, 2$ (see Lemma 3.18 for a verification that (3.78) is sufficient for the first inequality; the last inequality follows from the definition of ν_0, see (3.71)).

Recall that $\delta > 0$ was fixed in Lemma 3.7 (i.e. when we were defining h_1, h_2) and it also appears in the definition of q^k_1, q^k_2 (recall (3.74)).

Using (3.77) we obtain

$$\partial_t |u(x, 0, t)|^2 = \partial_t q_{1,t}^k(x)^2 + \partial_t q_{2,t}^k(x)^2$$

$$= -2\delta - \left(a_1^k(t)v_1(x) + a_2^k(t)v_2(x) \right) \cdot \nabla \left(h_{1,t}(x)^2 + h_{2,t}(x)^2 \right.$$

$$\left. + 2p[a_1^k(t)v_1, h_{1,t}](x) + 2p[a_2^k(t)v_2, h_{2,t}](x) \right)$$

$$\leq -\delta - \left(a_1^k(t)v_1(x) + a_2^k(t)v_2(x) \right) \cdot \nabla \left(q_{1,t}^k(x)^2 + q_{2,t}^k(x)^2 \right.$$

$$\left. + 2p[a_1^k(t)v_1, q_{1,t}^k](x) + 2p[a_2^k(t)v_2, q_{2,t}^k](x) \right)$$

$$= -\delta - u_1(x, 0, t)\partial_{x_1}\left(|u(x, 0, t)|^2 + 2p(x, 0, t) \right)$$

$$-u_2(x, 0, t)\partial_{x_2}\left(|u(x, 0, t)|^2 + 2p(x, 0, t) \right), \qquad (3.84)$$

and so, recalling that $\partial_{x_3}|u(x, 0, t)|^2 = \partial_{x_3}p(x, 0, t) = 0$ (as a property of axisymmetric functions, see (3.31) and (3.36)),

$$\partial_t |u(x, 0, t)|^2 \leq -\delta - u(x, 0, t) \cdot \nabla \left(|u(x, 0, t)|^2 + 2p(x, 0, t) \right)$$

$$\leq 2\nu u(x, 0, t) \cdot \Delta u(x, 0, t) - u(x, 0, t) \cdot \nabla \left(|u(x, 0, t)|^2 + 2p(x, 0, t) \right)$$

for all $\nu \in [0, \nu_0]$, where we used (3.83) in the last step.

Thus, we have shown that for sufficiently large k the Navier–Stokes inequality (3.73) holds for all $x \in P$, $t \in [0, T]$ and $\nu \in [0, \nu_0]$, which gives (iii), as required.

3.3.3 The Oscillatory Processes

Here we construct the oscillatory processes $a_1^k, a_2^k \in C^\infty(\mathbb{R}, [-1, 1])$, $k \geq 1$, such that the functions q_1^k, q_2^k (given by (3.74)) converge to h_1, h_2 (respectively) as in (3.78). As outlined in Sect. 3.3.1, this completes the proof of Theorem 3.1 given the *geometric arrangement* (which we construct in Sect. 3.4).

As for the strategy for choosing a_1^k, a_2^k we will divide $[0, T]$ into $4k$ subintervals and on each those subintervals we will let each of a_1^k, a_2^k equal 1, -1 or 0 (except for a set of times of measure less than $1/k$) in a particular configuration. The configuration is such that the resulting $q_{1,t}^k, q_{2,t}^k$ oscillate near $h_{1,t}, h_{2,t}$ as t varies between 0 and T, and such that the oscillations grow in frequency (that is the number of subintervals increases with k) and decrease in magnitude (that is we obtain convergence (3.78)), see Fig. 3.6 for a sketch.

We employ this strategy in the proof of the theorem below.

Theorem 3.9 (existence of the oscillatory processes) *For each $k \geq 1$, there exist a pair of functions $a_i^k \in C^\infty(\mathbb{R}; [-1, 1])$, $i = 1, 2$, such that*

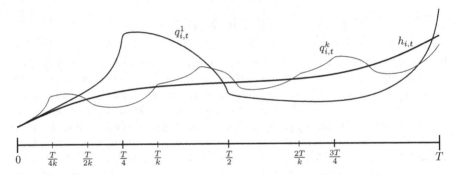

Fig. 3.6 The strategy for the choice of a_i^k, $i = 1, 2$. This sketch illustrates how the choice of a_i^k's causes $q_{i,t}^k$'s to "oscillate around $h_{i,t}$" as t varies between 0 to T. Here $k = 3$

$$\int_0^t a_i^k(s) \left(G_i(x, s) + F_{i,1}\left(x, s, a_1^k(s)\right) + F_{i,2}\left(x, s, a_2^k(s)\right)\right) ds$$

$$\xrightarrow{k \to \infty} \begin{cases} \frac{1}{2} \int_0^t \left(F_{2,1}(x, s, 1) - F_{2,1}(x, s, 0)\right) ds & i = 2, \\ 0 & i = 1 \end{cases} \tag{3.85}$$

uniformly in $(x, t) \in P \times [0, T]$ for any bounded and uniformly continuous functions

$$G_i \colon P \times [0, T] \to \mathbb{R}, \qquad F_{i,l} \colon P \times [0, T] \times [-1, 1] \to \mathbb{R},$$

$i, l = 1, 2$, *satisfying*

$$F_{i,l}(x, t, -1) = F_{i,l}(x, t, 1) \quad \text{for } x \in P, t \in [0, T], i, l = 1, 2.$$

Note that this theorem gives (3.78) simply by taking

$$G_i(x, t) := v_i(x) \cdot \nabla h_i(x, t)^2,$$
$$F_{i,l}(x, t, a) := 2 v_i(x) \cdot \nabla p[a v_l, h_{l,t}](x)$$

(recall $p[v, f] = p[-v, f]$ by Lemma 3.3 (i)), and so such $F_{i,l}$'s satisfy the requirement $F_{i,l}(x, t, -1) = F_{i,l}(x, t, 1)$ above) and by taking

$$G_i(x, t) := D^\alpha \left(v_i(x) \cdot \nabla h_i(x, t)^2\right),$$
$$F_{i,l}(x, t, a) := D^\alpha \left(2 v_i(x) \cdot \nabla p[a v_l, h_{l,t}](x)\right)$$

for any given multiindex $\alpha = (\alpha_1, \alpha_2)$.

Before proceeding to the proof of Theorem 3.9 we pause for a moment to comment on the meaning of the theorem and the convergence (3.78) in an informal manner. Recall that (3.74) includes terms of the form

$$2 \int_0^t a_i^k(s) v_i \cdot \nabla p[a_l^k(s)v_l, h_{l,s}] ds, \quad i, l \in \{1, 2\}.$$

Note that each of such terms represent, in a sense, an influence of the set U_l (together with the structure $(a_l^k(s)v_l, h_{l,s}, \phi_l)$) on the set U_i, namely, it vanishes outside U_i and it uses the non-local character of the pressure function $p[\cdot, \cdot]$ (that is the fact that the pressure function $p[a_l^k(s)v_l, h_{l,s}]$ does not vanish on U_i). Thus, we see from (3.85) that the role of the oscillatory processes a_1^k, a_2^k is to "select" only the influence of U_1 on U_2 as $k \to \infty$ (except for this the oscillatory behaviour of the processes makes the terms $\int_0^t a_i^k(s) v_i \cdot \nabla h_{i,s}^2$, $i = 1, 2$, vanish as $k \to \infty$). Note this is the desired behaviour since we want to show the convergence (3.78) and of the two functions h_1, h_2 only h_2 includes an influence from U_1 (recall (3.64), (3.65)). The construction of such oscillatory processes is clear from the following auxiliary considerations, in which we forget, for a moment, about the smoothness requirement.

Let $f : [-1, 1] \to \mathbb{R}$ be such that $f(-1) = f(1)$ and let functions $b_1, b_2 : [0, T] \to [-1, 1]$ be such that

$$b_1(t) = \begin{cases} 1 & t \in (0, T/4), \\ -1 & t \in (T/4, T/2), \\ 0 & t \in (T/2, T), \end{cases} \qquad b_2(t) = \begin{cases} 1 & t \in (0, T/2), \\ -1 & t \in (T/2, T). \end{cases} \tag{3.86}$$

Then

$$\int_0^T b_i(s) f(b_l(s)) \, ds = \begin{cases} \frac{T}{2}(f(1) - f(0)) & (i, l) = (2, 1), \\ 0 & (i, l) \neq (2, 1), \end{cases} \tag{3.87}$$

that is, the choice of b_1, b_2 is such that they "pick" the value $T(f(1) - f(0))/2$ only for the choice of indices $(i, l) = (2, 1)$. Clearly, given $(i_0, l_0) \in \{1, 2\}^2$ one could choose b_1, b_2 that pick this value only for the choice of indices $(i, l) = (i_0, l_0)$.

More generally, let f be also a function of time, $f : [0, T] \times [-1, 1] \to \mathbb{R}$ with $f(t, -1) = f(t, 1)$ for all t such that f is almost constant with respect to the first variable, i.e. for some $\epsilon > 0$

$$\sup_{t \in [0, T]} f(t, a) - \inf_{t \in [0, T]} f(t, a) < \epsilon, \qquad a \in [-1, 1].$$

Then

$$\int_0^T b_i(t) f(s, b_l(s)) \, ds = \begin{cases} \frac{1}{2} \int_0^T (f(s, 1) - f(s, 0)) \, ds + T \, O(\epsilon) & (i, l) = (2, 1), \\ T \, O(\epsilon) & (i, l) \neq (2, 1). \end{cases}$$

These observations are helpful in finding $b_1^k, b_2^k : [0, T] \to [-1, 1]$ such that for every continuous f

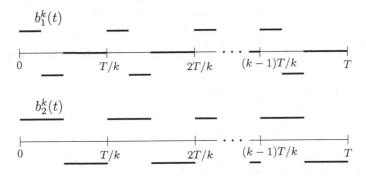

Fig. 3.7 The functions b_1^k, b_2^k

$$\int_0^T b_i^k(s) f\left(s, b_l^k(s)\right) ds \rightarrow \begin{cases} \frac{1}{2} \int_0^T (f(s, 1) - f(s, 0)) ds & (i, l) = (2, 1), \\ 0 & (i, l) \neq (2, 1) \end{cases} \quad (3.88)$$

as $k \rightarrow \infty$. Indeed one can take b_1^k, b_2^k to be oscillations of the form (3.86), but of higher frequency,

$$b_1^k(t) := b_1(kt), \qquad b_2^k(t) := b_2(kt), \quad (3.89)$$

where we extended b_1, b_2 T-periodically to the whole line, see Fig. 3.7.

In order to see that such a choice gives the convergence in (3.88) note that continuity of f implies that $\varepsilon_k \rightarrow 0$ as $k \rightarrow \infty$, where $\varepsilon_k > 0$ is the smallest positive number such that

$$|f(t, a) - f(s, a)| \leq \varepsilon_k \quad (3.90)$$

whenever $a \in [-1, 1]$ and $s, t \in [0, T]$ are such that $|t - s| \leq T/k$. Thus, if $(i, l) = (2, 1)$ we write

$$\int_0^T b_2^k(s) f\left(s, b_1^k(s)\right) ds = \sum_{p=0}^{k-1} \int_{pT/k}^{(p+1)T/k} b_2^k(s) f\left(s, b_1^k(s)\right) ds$$

$$= \sum_{p=0}^{k-1} \left(\int_{pT/k}^{(p+1/2)T/k} f(s, 1) \, ds - \int_{(p+1/2)T/k}^{(p+1)T/k} f(s, 0) \, ds \right)$$

$$= \sum_{p=0}^{k-1} \left(\frac{1}{2} \int_{pT/k}^{(p+1)T/k} f(s, 1) \, ds - \frac{1}{2} \int_{pT/k}^{(p+1)T/k} f(s, 0) \, ds + \frac{T}{k} O(\varepsilon_k) \right)$$

$$= \frac{1}{2} \int_0^T (f(s, 1) - f(s, 0)) ds + T \, O(\varepsilon_k).$$

$$(3.91)$$

Thus

$$\int_0^T b_2^k(s) f\left(s, b_1^k(s)\right) ds \to \frac{1}{2} \int_0^T (f(s, 1) - f(s, 0)) ds \qquad \text{as } k \to \infty,$$

and in the same way, one can show that

$$\int_0^T b_i^k(s) f\left(s, b_l^k(s)\right) ds \to 0 \qquad \text{as } k \to \infty$$

if $(i, l) \neq (2, 1)$. Therefore we obtain (3.88).

In a similar way, one can show that for such choice of b_1^k, b_2^k, the upper limit of the integrals in (3.88) can be replaced by any $t \in [0, T]$, that is,

$$\int_0^t b_i^k(s) f\left(s, b_j^k(s)\right) ds \to \begin{cases} \frac{1}{2} \int_0^t (f(s, 1) - f(s, 0)) ds & (i, l) = (2, 1), \\ 0 & (i, l) \neq (2, 1) \end{cases} \quad (3.92)$$

as $k \to \infty$, uniformly in $t \in [0, T]$. To this end, given $t \in [0, T]$ let $q \in \{0, \dots, k - 1\}$ be such that $t \in [qT/k, (q+1)T/k)$ and write the left-hand side of (3.92) above as

$$\sum_{p=0}^{q-1} \int_{pT/k}^{(p+1)T/k} b_2^k(s) f(s, b_1^k(s)) ds + \int_{qT/k}^t b_2^k(s) f(s, b_1^k(s)) ds.$$

The sum from $p = 0$ to $q - 1$ can be treated in the same way as the sum over all p's in the calculation (3.91) above to give

$$\frac{1}{2} \int_0^{qT/k} (f(s, 1) - f(s, 0)) ds + \frac{qT}{k} O(\varepsilon_k).$$

The remaining term can be treated using boundedness of f (note $|f| \le N$ for some $N > 0$ due to continuity of f and to the fact that its domain $[0, T] \times [-1, 1]$ is compact) by writing

$$\left| \int_{qT/k}^t b_2^k(s) f(s, b_1^k(s)) ds - \frac{1}{2} \int_{qT/k}^t (f(s, 1) - f(s, 0)) ds \right| \le 2N |t - qT/k| \le 2NT/k,$$

and thus we obtain (3.92) in the case $(i, l) = (2, 1)$. The case $(i, l) \neq (2, 1)$ follows similarly.

Moreover, due to the oscillatory behaviour of b_1^k, b_2^k as k increases we also see that each of b_1^k, b_2^k converges to 0 in a weak sense, that is,

$$\int_0^t b_i^k(s) g(s) ds \to 0 \qquad \text{as } k \to \infty, i = 1, 2, \text{ uniformly in } t \in [0, T] \quad (3.93)$$

for any continuous $g\colon [0, T] \to \mathbb{R}$.

The above ideas are a basis of the proof of Theorem 3.9, in which x plays no role and the processes a_1^k, a_2^k are obtained by a smooth approximation of b_1^k, b_2^k, respectively.

Proof of Theorem 3.9 Let $b_1^k, b_2^k\colon [0, T] \to [-1, 1]$ be defined by (3.89) above. Given $k \geq 0$ let $\varepsilon_k > 0$ be the smallest number such that

$$|F_{i,l}(x, t, a) - F_{i,l}(x, s, a)|, |G_i(x, t) - G_i(x, s)| \leq \varepsilon_k, \qquad i, l = 1, 2 \quad (3.94)$$

whenever $x \in P$, $a \in [-1, 1]$ and $t, s \in [0, T]$ are such that $|t - s| \leq T/k$. Due to the uniform continuity of $F_{i,l}$'s and G_i's we obtain $\varepsilon_k \to 0$ as $k \to \infty$. Moreover, from boundedness we obtain $N > 0$ such that $|F_{i,l}|, |G_i| \leq N$ for $i, j = 1, 2$. Thus, applying (3.92), with $f(t, a) := F_{i,l}(x, t, a)$ (for every x) and with the continuity property (3.90) replaced by the uniform continuity of $F_{i,j}$'s (3.94) and by the boundedness $|F_{i,l}| \leq N$ we obtain

$$\int_0^t b_i^k(s) F_{i,l}\left(x, s, b_l^k(s)\right) ds \to \begin{cases} \frac{1}{2} \int_0^t (F_{2,1}(x, s, 1) - F_{2,1}(x, s, 0)) ds & (i, l) = (2, 1), \\ 0 & (i, l) \neq (2, 1) \end{cases}$$

as $k \to \infty$ uniformly in $x \in P$, $t \in [0, T]$. Similarly, applying (3.93) with $g(t) := G_i(x, t)$ we obtain

$$\int_0^t b_i^k(s) G_i(x, s) \, ds \to 0 \qquad \text{as } k \to \infty$$

uniformly in $x \in P$, $t \in [0, T]$, $i = 1, 2$. Thus, altogether

$$\int_0^t b_i^k(s) \left(G_i(x, s) + F_{i,1}\left(x, s, b_1^k(s)\right) + F_{i,2}\left(x, s, b_2^k(s)\right)\right) ds$$

$$\xrightarrow{k \to \infty} \begin{cases} \frac{1}{2} \int_0^t \left(F_{2,1}(x, s, 1) - F_{2,1}(x, s, 0)\right) ds & i = 2, \\ 0 & i = 1 \end{cases} \qquad (3.95)$$

uniformly in $(x, t) \in P \times [0, T]$. Thus, the oscillatory processes b_1^k, b_2^k (defined by (3.89)) satisfy all the claims of the theorem, except for the C^∞ regularity. To this end let $a_1^k, a_2^k \in C^\infty(\mathbb{R}; [-1, 1])$ be such that

$$\left|\{t \in [0, T]\colon a_i^k(t) \neq b_i^k(t)\}\right| \leq \frac{1}{k}, \qquad i = 1, 2.$$

Such a_1^k, a_2^k can be obtained by extending b_1^k, b_2^k to the whole line by zero and mollifying. Clearly, such definition of the processes a_1^k, a_2^k and the boundedness $|F_{i,l}|, |G_i| \leq N$ gives that the difference between the left-hand sides of (3.85) and (3.95) is bounded by

$$6N/k \to 0 \qquad \text{as } k \to \infty,$$

which shows that these left-hand sides converge to the same limit

$$\begin{cases} \frac{1}{2} \int_0^t \left(F_{2,1}(x, s, 1) - F_{2,1}(x, s, 0) \right) ds & i = 2, \\ 0 & i = 1 \end{cases}$$

uniformly in $(x, t) \in P \times [0, T]$, as required.   □

3.4 The Geometric Arrangement

In this section, we construct the *geometric arrangement*, that is, $T > 0$, $\tau \in (0, 1)$, $z \in \mathbb{R}^3$, sets $U_1, U_2 \Subset P$ with disjoint closures and the respective structures (v_1, f_1, ϕ_1), (v_2, f_2, ϕ_2) such that

$$f_2^2 + T v_2 \cdot F[v_1, f_1] > |v_2|^2 \quad \text{in } U_2,$$

$$f_2^2(y) + T v_2(y) \cdot F[v_1, f_1](y) > \tau^{-2} \left(f_1(R^{-1}x) + f_2(R^{-1}x) \right)^2$$

for all $x \in G = R(\overline{U_1} \cup \overline{U_2})$, where $y = R^{-1}(\Gamma(x))$. According to the considerations of Sect. 3.3, this construction concludes the proof of Theorem 3.1.

Let

$$U := (-1, 1) \times (1/8, 7/8),$$

and let $v \in C_0^\infty(U; \mathbb{R}^2)$ be any vector field satisfying

$$\begin{cases} v_1(-x_1, x_2) = v_1(x_1, x_2), \\ v_2(-x_1, x_2) = -v_2(x_1, x_2) \end{cases}$$

and

$$\operatorname{div}(x_2 \, v(x_1, x_2)) = 0, \qquad (x_1, x_2) \in P.$$

One can take, for instance,

$$v(x_1, x_2) := x_2^{-1} J \left((-(x_2 - 1/2), x_1) \chi_{\{1/16 < |(x_1, x_2 - 1/2)| < 1/8\}} \right),$$

where J denotes a sufficiently fine mollification (and χ denotes the indicator function), as in the recipe for a structure presented in Sect. 3.2.5. Following the recipe, let $f \in C_0^\infty(P; [0, \infty))$ be such that supp $f = \overline{U}$, $f > |v|$ in U and $Lf > 0$ at points of U of sufficiently small distance from ∂U. Furthermore, construct f in a way that

$$f(-x_1, x_2) = f(x_1, x_2).$$

We show existence of such f in Lemma 3.15. Let $\phi \in C_0^\infty(U; [0, 1])$ be a cut-off function such that supp $v \subset \{\phi = 1\}$ and $Lf > 0$ in $U \setminus \{\phi = 1\}$. Thus, we obtained a structure (v, f, ϕ) on U. Consider the pressure interaction function $F := F[v, f]$ (recall (3.48)) and let $A \in \mathbb{R}$, $B, C, D, N > 0$ and $\kappa = 10^4 C/D$ be the constants given by Lemma 3.6.

Since the structure (v, f, ϕ) satisfies the condition of Lemma 3.3 (ii), we see that the first component of $F[v, f]$ is odd when restricted to the x_1 axis, that is,

$$F_1(-x_1, 0) = -F_1(x_1, 0), \quad x_1 \in \mathbb{R}.$$

Thus, in the view of Lemma 3.6 (ii), we observe that $A \neq 0$ and

$$-B = F_1(-A, 0) = \min_{x_1 \in \mathbb{R}} F_1(x_1, 0).$$

3.4.1 A Simplified Geometric Arrangement

At this point, we pause for the moment to present a certain simplified geometric arrangement. Although the simplified arrangement has the unfortunate property of being impossible, it offers a good perspective on the main difficulty. We also explain the strategy for overcoming this difficulty. The reader who is not interested in the simplified arrangement is referred to the Sect. 3.4.2, where we proceed with the presentation of the geometric arrangement proper.

From Lemma 3.6 (ii), we see that there exists a rectangle $U_2 \Subset P$ such that $F_1[v, f] \geq B/2$ in U_2. Let $v_2 = (v_{21}, v_{22}) \in C_0^\infty(U_2; \mathbb{R}^2)$ be such that div $(x_2 v_2 (x_1, x_2)) = 0$ for $(x_1, x_2) \in P$,

$$v_{22} = 0, \quad v_{21} \geq 0 \text{ and } v_2 = (1, 0) \text{ in some closed rectangle } K \subset U_2.$$

Warning 3.10 *Such v_2 does not exist!*

Indeed, take $w := x_2 v_2$ and let K' be a rectangle such that its left edge is the left edge of K and its right edge lies on ∂U_2. Integrating div *w over K we obtain*

$$0 = \int_K \text{div } w = \int_{\partial K'} w \cdot n = \int_{\partial_L K'} w_1 = \int_{\partial_L K'} x_2 > 0,$$

where $\partial_L K'$ denotes the left edge of K'.

Let (z_1, z_2) be an interior point of K, $z := (z_1, z_2, 0) \in \mathbb{R}^3$, $U_1 := U$, $v_1 := v$, $f_1 := f$, $\phi_1 := \phi$ (note then $F = F[v_1, f_1]$) and let $\tau \in (0, 1)$ be sufficiently small such that

$$R^{-1}\left(\tau R(\overline{U_1} \cup \overline{U_2}) + z\right) \subset K, \tag{3.96}$$

Fig. 3.8 The simplified
arrangement. Note that it is
not quite correct, see
Warning 3.10

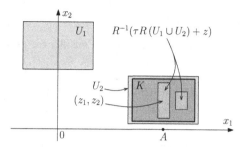

see Fig. 3.8. Let f_2, ϕ_2 be any functions such that (v_2, f_2, ϕ_2) is a structure on U_2
(that is define f_2, ϕ as described in the recipe in Sect. 3.2.5). Then (3.51) follows
trivially for every $T > 0$ by noting that

$$v_2 \cdot F = v_{21} F_1 \geq 0 \quad \text{in } U_2, \tag{3.97}$$

and so

$$f_2^2 + T v_2 \cdot F = f_2^2 + T v_{21} F_1 \geq f_2^2 > |v_2|^2 \quad \text{in } U_2. \tag{3.98}$$

Moreover, (3.52) follows provided we choose $T > 2\tau^{-2} \|f_1 + f_2\|_\infty^2 / B$. Indeed, then
we obtain

$$T F_1 \geq \tau^{-2} \|f_1 + f_2\|_\infty^2 \quad \text{in } U_2, \tag{3.99}$$

and so letting $x \in R(\overline{U_1} \cup \overline{U_2})$ and $y := R^{-1}(\tau x + z)$ we see that (3.96) gives $y \in K$
and thus

$$f_2^2(y) + T v_2(y) \cdot F(y) = f_2^2(y) + T F_1(y) \geq \tau^{-2} \|f_1 + f_2\|_\infty^2, \tag{3.100}$$

as required.

This concludes the simplified geometric arrangement. Note, however, it does not
exist due to Warning 3.10. In fact, it is clear that v_2 cannot have $(1, 0)$ as the only
direction, which is, roughly speaking, a consequence of the fact that any weakly
divergence-free vector field in $\mathbb{R}^2$ must "run in a loop", cf. Fig 3.3. Thus, given any
of the quantities

$$F_1, F_2, -F_1, -F_2$$

there exists a region in P such that at least one of the ingredients of the inner product

$$v_2 \cdot F = v_{21} F_1 + v_{22} F_2$$

gives the given quantity multiplied by v_{21} or v_{22} (the size of which obviously depending on the choice of v_2). Thus, the calculations (3.98), (3.100), in which we used
the very convenient properties (3.97), (3.99) immediately become useless and at this
point it is not clear how to estimate $v_2 \cdot F$ to obtain the required relations (3.51),
(3.52).

In the remainder of this section, we sketch a more elaborate construction of sets U_1 and U_2 as well as their structures that solve this difficulty. In particular, we point out the relations that will replace (3.97), (3.99) in showing the required relations (3.51), (3.52). The construction is then presented in detail in the following Sects. 3.4.2–3.4.5.

First of all, we will consider the rescaling of the set U and its structure (v, f, ϕ), that is, for $\alpha \in \mathbb{R}$, $\rho > 0$ and $\sigma > 0$ we will consider a set $U^{\alpha,\rho}$ and a structure $(v^{\alpha,\rho,\sigma}, f^{\alpha,\rho,\sigma}, \phi^{\alpha,\rho})$ on $U^{\alpha,\rho}$. Here α corresponds to a translation in the x_1 direction, ρ scales the size of U and σ scales the magnitude of v and f. We will observe that manipulating the values of α, ρ, σ gives us certain amount of freedom in the manipulation of the shape of the pressure interaction function

$$F^{\alpha,\rho,\sigma} := F[v^{\alpha,\rho,\sigma}, f^{\alpha,\rho,\sigma}],$$

and so we will consider a disjoint union of U together with its two rescalings,

$$U \cup U^{a',r'} \cup U^{a'',r''},$$

along with the corresponding structure

$$(v, f, \phi) + \left(v^{a',r',s'}, f^{a',r',s'}, \phi^{a',r'} \right) + \left(v^{a'',r'',s''}, f^{a'',r'',s''}, \phi^{a'',r''} \right),$$

where the sum is understood in an entry-wise sense. Here, the values of a', a'', r', r'', s', s'' will be chosen in a particular way, roughly speaking such that the (joint) pressure interaction function

$$H := F + F^{a',r',s'} + F^{a'',r'',s''}$$

enjoys a similar decay to F (recall Lemma 3.6 (iii)) and, when restricted to the x_1 axis, its first component H_1 admits maximum $7B$ at A and minimum greater than or equal to $-1.005B$ (rather than maximum B and minimum $-B$, which is the case for F_1), see Fig. 3.9. Then, given a small parameter $\varepsilon > 0$, we will find numbers $d, r = O(1/\varepsilon)$ with $d \gg r$, $U_2 \Subset P$ and $v_2 \in C_0^\infty(U_2; \mathbb{R}^2)$ such that

$$U_2 \subset BOX := [-d, d] \times [0, r],$$

U_2 is a rectangular ring encompassing $U \cup U^{a',r'} \cup U^{a'',r''}$,

namely, $U_2 = V \setminus \overline{W}$ where $V, W \Subset P$ are rectangles such that

$$U \cup U^{a',r'} \cup U^{a'',r''} \Subset W \Subset V,$$

see Fig. 3.9, and

$$v_2 = (1, 0) \quad in \ RECT \subset U_2,$$

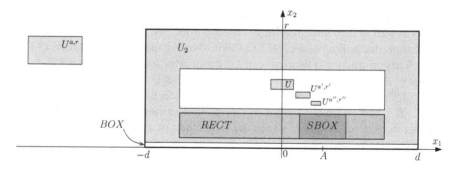

Fig. 3.9 A sketch of the geometric arrangement (see Fig. 3.14 for a more detailed sketch). Some proportions are not conserved on the sketch

where $RECT$ will be a carefully chosen rectangle located sufficiently close to the x_1 axis so that

$$H_1 \geq -1.01B \text{ in } RECT,$$
$$H_1 \geq 6.99B \text{ in some rectangle } SBOX \subset RECT. \tag{3.101}$$

We will then choose τ, z such that

$$R^{-1}(\tau R(BOX) + z) \subset SBOX, \tag{3.102}$$

see Fig. 3.14, and we will define a pair of numbers $a = O(-\varepsilon^{-2}), s = O(\varepsilon^{-5/2})$ such that the rescaling $U^{a,r}$ of U together with the rescaled structure $(v^{a,r,s}, f^{a,r,s}, \phi^{a,r})$ satisfies

$$R^{-1}\left(\tau R\left(\overline{U^{a,r}}\right) + z\right) \subset RECT, \tag{3.103}$$

see Fig. 3.14, and that the pressure interaction function $F^{a,r,s} = F[v^{a,r,s}, f^{a,r,s}]$ is of particular size when restricted to BOX, that is, $F_2^{a,r,s}$ is small (in some sense) and

$$1.03B \leq F_1^{a,r,s} \leq 1.05B \quad \text{in } BOX. \tag{3.104}$$

For this, we will crucially need the last property in Lemma 3.6, which, roughly speaking, quantifies the decay (in x_1) of the pressure interaction function. We will then set

$$U_1 := U \cup U^{a',r'} \cup U^{a'',r''} \cup U^{a,r}$$

together with the structure

$$(v_1, f_1, \phi_1) := (v, f, \phi) + \left(v^{a',r',s'}, f^{a',r',s'}, \phi^{a',r'}\right)$$
$$+ \left(v^{a'',r'',s''}, f^{a'',r'',s''}, \phi^{a'',r''}\right) + \left(v^{a,r,s}, f^{a,r,s}, \phi^{a,r}\right),$$

so that the (total) pressure interaction function is

$$F^* := F[v_1, f_1] = H + F^{a,r,s}.$$

Observe that (3.102), (3.103) give, in particular,

$$R^{-1}\left(\tau R\left(\overline{U_1} \cup \overline{U_2}\right) + z\right) \subset RECT,$$

that is, as in the simplified setting (see (3.96)), the cylindrical projection R^{-1} maps $\Gamma(G)$ (recall $G = R\left(\overline{U_1} \cup \overline{U_2}\right)$) into the region in P in which $v_2 = (1, 0)$. Moreover, (3.101) and (3.104) immediately give

$$\begin{aligned} F_1^* &> 0.01\,B \quad \text{in } RECT, \\ F_1^* &> 8B \quad\;\; \text{in } SBOX. \end{aligned} \tag{3.105}$$

Furthermore, it can be shown (using the properties of the choice of $\varepsilon, d, r, a, s, v_2$ and the decay of H) that

$$v_2 \cdot F^* \geq -1.1\varepsilon B \quad \text{in supp } v_2. \tag{3.106}$$

Finally, we will make a particular choice of f_2, ϕ_2 and $T > 0$ such that (v_2, f_2, ϕ_2) is a structure on U_2 and the properties (3.51), (3.52) hold. The proof of (3.51) will be in essence similar to the calculation (3.98), but with the inequality (3.97) replaced by (3.106) and a property of the choice of T. The proof of (3.52) is, in a sense, a more elaborate version of the calculation (3.100). Namely, rather than taking any $x \in R\left(\overline{U_1} \cup \overline{U_2}\right)$ we will consider two cases, which correspond to different means of substituting the use of the inequality (3.99):

$Case\ 1.\ x \in R\left(\overline{U^{a,r}}\right)$. Then $y \in RECT$ by (3.103) and we will replace (3.99) by the first inequality in (3.105) and the properties of f_2 and T.

$Case\ 2.\ x \in R\left(\overline{U} \cup \overline{U^{a',r'}} \cup \overline{U^{a'',r''}} \cup \overline{U_2}\right) \subset R(BOX)$. Then $y \in SBOX$ by (3.102) and we will replace (3.99) by the second inequality in (3.105) and the properties of f_2 and T.

We now present the rigorous version of this explanation.

3.4.2 The Copies of U and Its Structure

Let us consider disjoint "copies" of U and its structure (v, f, ϕ) and arranging these copies into a favourable composition. Namely, for $\alpha \in \mathbb{R}, \rho > 0, \sigma > 0$ let

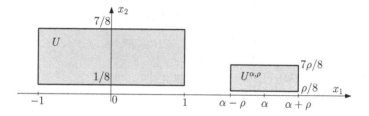

Fig. 3.10 The set $U^{\alpha,\rho}$, where $\rho < 1$

$$U^{\alpha,\rho} := \left\{ (x_1, x_2) \in \mathbb{R}^2 : \left(\frac{x_1 - \alpha}{\rho}, \frac{x_2}{\rho} \right) \in U \right\},$$

$$v^{\alpha,\rho,\sigma}(x_1, x_2) := \sigma \, v \left(\frac{x_1 - \alpha}{\rho}, \frac{x_2}{\rho} \right),$$

$$f^{\alpha,\rho,\sigma}(x_1, x_2) := \sigma \, f \left(\frac{x_1 - \alpha}{\rho}, \frac{x_2}{\rho} \right), \tag{3.107}$$

$$\phi^{\alpha,\rho}(x_1, x_2) := \phi \left(\frac{x_1 - \alpha}{\rho}, \frac{x_2}{\rho} \right),$$

$$F^{\alpha,\rho,\sigma}(x_1, x_2) := \frac{\sigma^2}{\rho} F \left(\frac{x_1 - \alpha}{\rho}, \frac{x_2}{\rho} \right).$$

(Recall $F = F[v, f]$ is the pressure interaction function.)

Here $\alpha \in \mathbb{R}$ denote the translation in x_1 direction of U and its structure and ρ denotes the scaling of the variables, see Fig. 3.10. Also, σ denotes the scaling in magnitude of v and f. A direct consequence of the definitions above is that $U^{\alpha,\rho} \in P$, $(v^{\alpha,\rho,\sigma}, f^{\alpha,\rho,\sigma}, \phi^{\alpha,\rho})$ is a structure on $U^{\alpha,\sigma}$ and $F^{\alpha,\rho,\sigma}$ is a pressure interaction function corresponding to $U^{\alpha,\sigma}$, namely,

$$F^{\alpha,\rho,\sigma} = F[v^{\alpha,\rho,\sigma}, f^{\alpha,\rho,\sigma}],$$

for each choice of $\alpha \in \mathbb{R}$, $\rho, \sigma > 0$. Now let $a', a'' \in \mathbb{R}$, $r', r'', s', s'' > 0$ be such that the sets U, $U^{a',r'}$, $U^{a'',r''}$ have disjoint closures and the function

$$H := F + F^{a',r',s'} + F^{a'',r'',s''} \tag{3.108}$$

satisfies

(i) $H_1(A, 0) = 7B$,
(ii) $H_1(x_1, 0) \geq -1.005 B$,
(iii) $|H(x)| \leq 2C/|x|^4$ for $|x| > 2|A|$.

Such a choice is possible due to the following simple geometric argument (which is sketched in Fig. 3.11). Let s', r' satisfy $(s')^2/r' = 2$ (so that we have max $F_1^{a',r',s'}$ $(\cdot, 0) = 2B = - \min F_1^{a',r',s'}(\cdot, 0)$) and take $r' > 0$ so small that $|F_1^{0,r',s'}(x_1, 0)| <$

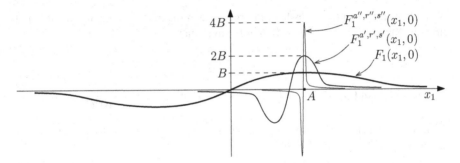

Fig. 3.11 The choice of $a', a'', r', r'', s', s''$

$0.001B$ for x_1 such that $F_1(A + x_1, 0) < 0.999B$. Then choose a' such that the maxima of both $F_1(x_1, 0)$ and $F_1^{a', r', s'}(x_1, 0)$ coincide (at $x_1 = A$). Then, similarly, choose s'', r'' so that $(s'')^2/r'' = 4$ and $r'' > 0$ is small enough so that $|F_1^{0, r'', s''}(x_1, 0)| < 0.001B$ for x_1 such that $F_1^{a', r', s'}(A + x_1, 0) < 0.999 \cdot (2B)$, and choose a'' so that the maximum of $F_1^{a'', r'', s''}(x_1, 0)$ occurs at $x_1 = A$. This way we obtain (i) and (ii) by construction, while (iii) follows given r' and r'' were chosen small enough. Furthermore, taking r' and r'' small ensures that the sets U, $U^{a', r'}$, $U^{a'', r''}$ have disjoint closures ($r' < 1/8$ and $r'' < r'/8$ suffices, cf. Fig. 3.9).

Thus, by specifying $a', a'', r', r'', s', s''$ we added to U two disjoint copies of it such that the total pressure interaction function H has a specific behaviour on the x_1 axis. We now want to specify the behaviour of H on a strip in P near the x_1 axis. That is, by continuity, we see that there exists $E > 0$ (sufficiently small) such that

(iv) the strip $\{0 < x_2 < E\} \subset P$ is disjoint from $U, U^{a', r'}, U^{a'', r''}$,
 (v) $H(x) \geq -1.01B$ in the strip $\{0 < x_2 < E\}$,
(vi) $H(x) \geq 6.99B$ for $x \in P$ such that $|x_1 - A| < \kappa E, 0 < x_2 < E$.

Here claim (v) also uses the decay property (iii) of H.

3.4.3 Construction of v_2 and U_2

Now let $\varepsilon > 0$ be a small parameter (whose value we fix below) and let $d, r > 0$ be defined by

$$r := E/\varepsilon, \quad d := \kappa r.$$

Note that by taking ε small, both r and d become large, and since

$$\kappa = 10^4 C/D \geq 10^4 \quad \text{we have} \quad d \geq 10^4 r. \tag{3.109}$$

In fact, ε is the main parameter of the construction and in what follows we will use certain algebraic inequalities, all of which rely on ε being sufficiently small.

We gather all these properties here in order to demonstrate that the argument is not circular. Namely, let $\varepsilon > 0$ be sufficiently small that

$$\varepsilon < 1/10, \quad d - r > 2(|A| + \kappa E), \quad r > 10, \quad r > 20|A|,$$

$$d > 2 \operatorname{diam}\left(U \cup U^{a',r'} \cup U^{a'',r''}\right), \quad \varepsilon < \kappa/N, \quad \varepsilon^2 < \frac{BE^4}{2 \cdot 10^6 C}. \tag{3.110}$$

We now construct v_2 by sharpening the observation from Fig. 3.4. Namely, we let v_2 be given by the following lemma.

Lemma 3.11 *Given $d, r, \varepsilon > 0$ such that $d > r$, $\varepsilon < 1/10$ there exists $v_2 = (v_{21}, v_{22}) \in C_0^\infty(P; \mathbb{R}^2)$ such that*

(i) $\operatorname{div}(x_2 v_2(x_1, x_2)) = 0$,
(ii) $\operatorname{supp} v_2 \subset (-d, d) \times (0.005\varepsilon r, r) \setminus [-(d-r), d-r] \times [\varepsilon r, r/10]$,
(iii) $|v_{22}| < \varepsilon/2$, $-\varepsilon^2 \le v_{21} \le 1$ with

$$v_{21} \ge 0, v_{22} = 0 \quad in \quad [-(d-r), d-r] \times (0, \varepsilon r),$$

(iv) $v_2 = (1, 0)$ in $[-(d-r), d-r] \times [0.02\varepsilon r, 0.98\varepsilon r]$.

Before proving the lemma, we note that the construction of such a vector field v_2 is one of the central ideas of the proof of Theorem 3.1. We will shortly see that it is thanks to v_2 that we can overcome the difficulty posed by Warning 3.10. Indeed, we can already see (in part (iv) above) that v_2 keeps constant direction and magnitude in a rectangular-shaped subset of P which is located near the Ox_1 axis, and that $v_2 = O(\varepsilon)$ whenever its direction is different (which we will see in the proof below).

Proof Let $w: P \to \mathbb{R}^2$ be defined by

$$w(x_1, x_2) = \begin{cases} (x_2, 0) & \text{in } R_1, \\ \frac{\varepsilon}{2}(d - x_1, x_2) & \text{in } R_2, \\ -\varepsilon^2(x_2, 0) & \text{in } R_3, \\ \frac{\varepsilon}{2}(x_1 + d, -x_2) & \text{in } R_4, \\ 0 & \text{in } P \setminus (R_1 \cup R_2 \cup R_3 \cup R_4), \end{cases}$$

where regions R_1, R_2, R_3 and R_4 are as indicated in Fig. 3.12. Observe that these regions, and the form of w inside each of them, are defined in the way that w is divergence free inside each region and $w \cdot n$ is continuous across the boundary between any pair of neighbouring regions, where n denotes the unit normal vector of the boundary. Recall (from a recipe for a structure, Sect. 3.2.5) that this is sufficient for w to be weakly divergence free on $\mathbb{R}^2$. Therefore (as in the recipe for a structure, see Sect. 3.2.5), Jw is divergence free, smooth and compactly supported vector field on P, where J denotes any mollification operator. Thus, letting

$$v_2 = Jw/x_2$$

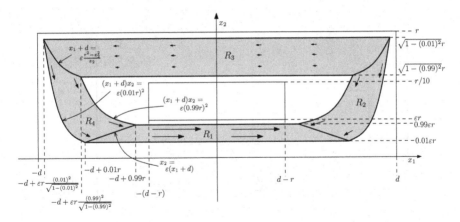

Fig. 3.12 The construction of w. The configuration of the curves in $\{x_1 > 0\}$ is defined symmetrically with respect to the x_2 axis. The arrows (inside the grey region) indicate the direction and magnitude of w. Note that some proportions are not conserved on this sketch

we see that, for sufficiently fine mollification J, v_2 satisfies all the required properties. In particular, $v_2 = (1, 0)$ in $[-(d-r), d-r] \times [0.02\varepsilon r, 0.98\varepsilon r]$ since affine functions are invariant under mollifications. □

Now let

$$\tau := 0.48\varepsilon, \qquad z := (A, \varepsilon r/2, 0). \qquad (3.111)$$

We see that

$$\tau d = \tau \kappa E / \varepsilon < \kappa E. \qquad (3.112)$$

Let

$$
\begin{aligned}
U_2 &:= (-d, d) \times (0.005\varepsilon r, r) \setminus [-(d-r), d-r] \times [\varepsilon r, r/10], \\
BOX &:= [-d, d] \times [0, r], \\
SBOX &:= [A - \kappa E, A + \kappa E] \times [0.02\varepsilon r, 0.98\varepsilon r], \\
RECT &:= [-(d-r), d-r] \times [0.02\varepsilon r, 0.98\varepsilon r],
\end{aligned}
\qquad (3.113)
$$

see Fig. 3.13.

Note that supp $v_2 \subset U_2$ by construction (see Lemma 3.11 (ii)) and that $SBOX \subset RECT$ by the second inequality in (3.110). Moreover,

$$R^{-1}(\tau R(BOX) + z) \subset SBOX. \qquad (3.114)$$

Indeed, since $\tau r < \varepsilon r/2$ we observe that the set on the left-hand side is simply

$$[A - \tau d, A + \tau d] \times [\varepsilon r/2 - \tau r, \varepsilon r/2 + \tau r] = [A - \tau d, A + \tau d] \times [0.02\varepsilon r, 0.98\varepsilon r] \subset SBOX,$$

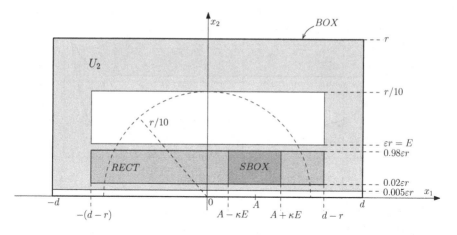

Fig. 3.13 The sets $U_2, BOX, RECT$ and $SBOX$. Note that proportions are not conserved on this sketch

where the inclusion follows from (3.112). What is more, the sets $U, U^{a',r'}, U^{a'',r''}$ are "encompassed" by U_2, that is,

$$U \cup U^{a',r'} \cup U^{a'',r''} \subset (-(d-r), d-r) \times (\varepsilon r, r/10), \qquad (3.115)$$

see Fig. 3.14. This property is clear from the identity $\varepsilon r = E$ and property (iv) of the choice of E (so that the strip $\{0 < x_2 < \varepsilon r\}$ is "below" these sets), the third inequality in (3.110) (so that the half-plane $\{x_2 > r/10\}$ is above U), and the fifth inequality in (3.110), which gives

$$d - r > \operatorname{diam}\left(U \cup U^{a',r'} \cup U^{a'',r''}\right)$$

(so that the length in the x_1 direction of the set $U \cup U^{a',r'} \cup U^{a'',r''}$ is less than $d - r$; recall also $d > 2r$ by (3.109)). Furthermore, properties (v) and (vi) of H (and the trivial inequality $0.98\varepsilon r \le E$) immediately give that

$$\begin{cases} H_1(x) \ge -1.01B & \text{in } (-(d-r), d-r) \times (0, \varepsilon r) \supset RECT, \\ H_1(x) \ge 6.99B & \text{in } SBOX. \end{cases} \qquad (3.116)$$

3.4.4 Construction of U_1 and Its Structure

We will add one more copy of U (and its structure) to the collection $U, U^{a',r'}, U^{a'',r''}$ (and the corresponding collection of structures). Namely, let

$$a := -\kappa r/\varepsilon, \quad \frac{s^2}{r} := 1.04\left(-\frac{a}{r}\right)^4 B/D, \qquad (3.117)$$

and consider $U^{a,r}$ with structure $(v^{a,r,s}, f^{a,r,s}, \phi^{a,r})$. In this way, the pressure inter-action function

$$F^{a,r,s} = F[v^{a,r,s}, f^{a,r,s}]$$

is of particular size in the whole of BOX, which we make precise in the following lemma.

Lemma 3.12

$$1.03B \leq F_1^{a,r,s} \leq 1.05B \quad and \quad |F_2^{a,r,s}| \leq 0.01\varepsilon B \quad in\ BOX.$$

Proof As for $F_1^{a,r,s}$ let $n := -a/r = \kappa/\varepsilon$ and observe that the sixth inequality in (3.110) gives $n \geq N$. Thus, since $|x_2|/r \leq 1$ and

$$\left|\frac{x_1 - a}{r} - n\right| = \frac{|x_1|}{r} \leq \frac{d}{r} = \kappa.$$

Lemma 3.6 (v) gives

$$\left|F_1\left(\frac{x_1 - a}{r}, \frac{x_2}{r}\right) - n^{-4}D\right| \leq 0.001 n^{-4} D.$$

(Recall (from the paragraph preceding Sect. 3.4.1) that $F = (F_1, F_2)$ denotes the pressure interaction function corresponding to U and structure (v, f, ϕ), that is, $F = F[v, f]$.) Therefore, since

$$F_1\left(\frac{x_1 - a}{r}, \frac{x_2}{r}\right) = \frac{r}{s^2} F_1^{a,r,s}(x_1, x_2) = \frac{n^{-4}D}{1.04B} F_1^{a,r,s}(x_1, x_2)$$

(recall (3.107) and (3.117)), we can multiply the last inequality by $1.04B/(n^{-4}D)$ to obtain

$$\left|F_1^{a,r,s}(x) - 1.04B\right| \leq 0.001(1.04B) < 0.01B \quad for\ x \in BOX.$$

As for $F_2^{a,r,s}$ let $(x_1, x_2) \in BOX$ and use Lemma 3.6 (iv), the Mean Value Theorem and Lemma 3.6 (iii) to write

$$\frac{r}{s^2}\left|F_2^{a,r,s}(x_1, x_2)\right| = \left|F_2\left(\frac{x_1 - a}{r}, \frac{x_2}{r}\right) - F_2\left(\frac{x_1 - a}{r}, 0\right)\right|$$

$$\leq \left|\nabla F_2\left(\frac{x_1 - a}{r}, \xi\right)\right| \left|\frac{x_2}{r}\right| \leq C\left|\frac{x_1 - a}{r}\right|^{-5},$$

where $\xi \in (0, 1)$. Thus, since the triangle inequality and the fact $\varepsilon < 1/2$ give

$$\frac{|x_1 - a|}{r} \geq \frac{|a|}{r} - \frac{|x_1|}{r} \geq \frac{|a|}{r} - \frac{d}{r} = \kappa \left(\frac{1}{\varepsilon} - 1 \right) \geq \frac{\kappa}{2\varepsilon},$$

we obtain (recalling (3.117) and that $\kappa = 10^4 C/D$, see (3.109))

$$\left| F_2^{a,r,s}(x_1, x_2) \right| \leq \left(2^5 \frac{1.04C}{D\kappa} \right) \varepsilon B = \frac{32 \cdot 1.04}{10^4} \varepsilon B < 0.01 \varepsilon B. \qquad \Box$$

Thus, letting

$$U_1 := U \cup U^{a',r'} \cup U^{a'',r''} \cup U^{a,r},$$
$$f_1 := f + f^{a',r',s'} + f^{a'',r'',s''} + f^{a,r,s},$$
$$v_1 := v + v^{a',r',s'} + v^{a'',r'',s''} + v^{a,r,s},$$
$$\phi_1 := \phi + \phi^{a',r'} + \phi^{a'',r''} + \phi^{a,r}$$

we obtain a structure (v_1, f_1, ϕ_1) on U_1, and denoting by F^* the total pressure interaction function,

$$F^* := F[v_1, f_1] = F + F^{a',r',s'} + F^{a'',r'',s''} + F^{a,r,s} = H + F^{a,r,s},$$

we see that the above lemma and (3.116) give

$$\begin{cases} F_1^* \geq 0.01B & \text{in } (-(d-r), d-r) \times (0, \varepsilon r) \supset RECT, \\ F_1^* \geq 8B & \text{in } SBOX. \end{cases} \qquad (3.118)$$

Moreover, the properties of H (the "joint" pressure interaction function of U, $U^{a',r'}$ and $U^{a'',r''}$, recall (3.108)), v_2, the smallness of ε (recall (3.110)) and the lemma above give

$$v_2 \cdot F^* \geq -1.1\varepsilon B \qquad \text{in } BOX, \qquad (3.119)$$

which we now verify. The claim for $x \in BOX \setminus \text{supp } v_2$ follows trivially. For $x \in \text{supp } v_2$, consider two cases.

Case 1. $|x| < r/10$. In this case, observe that since $d \geq 10^4 r$ (see (3.109)) we have $d - r > r/10$ and so $x \in (-(d-r), d-r) \times (0, \varepsilon r)$ (cf. Fig. 3.13). Thus, $v_{21}(x) \geq 0$, $v_{22}(x) = 0$ by construction of v_2 (see Lemma 3.11 (iii)), and so (3.118) gives

$$v_2(x) \cdot F^*(x) = v_{21}(x) F_1^*(x) \geq 0.01B v_{21}(x) \geq 0 > -1.1\varepsilon B.$$

Case 2. $|x| \geq r/10$. Since $r > 20|A|$ (see the fourth inequality in (3.110)), in this case $|x| \geq 2|A|$, and so property (iii) of H and the last inequality in (3.110) give

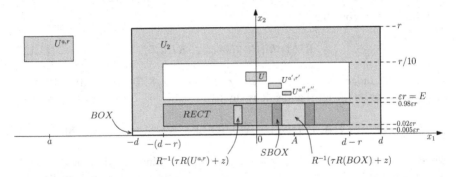

Fig. 3.14 The geometric arrangement (cf. Fig. 3.5)

$$|H(x)| \leq 2C/|x|^4 \leq 2C \left(\frac{10}{r}\right)^4 = 2 \cdot 10^4 C \varepsilon^4 / E^4 < 0.01 \varepsilon^2 B. \qquad (3.120)$$

This, the properties $-\varepsilon^2 \leq v_{21} \leq 1$, $|v_{22}| < \varepsilon/2$ (see Lemma 3.11 (iii)) and Lemma 3.12 give

$$\begin{aligned}
v_2(x) \cdot F^*(x) &= v_2(x) \cdot H(x) + v_{21}(x) F_1^{a,r,s}(x) + v_{22}(x) F_2^{a,r,s}(x) \\
&\geq -2(0.01 \varepsilon^2 B) - \varepsilon^2 (1.05 B) - \frac{\varepsilon}{2} (0.01 B \varepsilon) \\
&= -\varepsilon^2 B(0.02 + 1.05 + 0.005) \geq -1.1 \varepsilon^2 B.
\end{aligned}$$

Thus we obtain (3.119), as required.

Moreover, since $U^{a,r} = (a - r, a + r) \times (r/8, 7r/8)$, we see that

$$U^{a,r} \text{ is located "to the left" of } BOX, \qquad (3.121)$$

that is, $a + r < -d$ (see Fig. 3.14), which can be verified as follows. Since $\varepsilon < 1/10$ (recall (3.110)) and $\kappa > 1$ (recall (3.109)), we trivially obtain

$$\kappa \left(\frac{1}{\varepsilon} - 1\right) > 1,$$

which, multiplied by r, gives

$$-\frac{\kappa r}{\varepsilon} + r < -\kappa r,$$

that is, $a + r < -d$, as required. Thus, taking into account (3.115) we see that $\overline{U_1}$ and $\overline{U_2}$ are disjoint (see Fig. 3.14), which is one of the requirements of the geometric arrangement.

Furthermore, note that

$$R^{-1}\left(\tau R\left(\overline{U^{a,r}}\right)+z\right)\subset RECT, \tag{3.122}$$

see Fig. 3.14. Indeed, since $U^{a,r}=(a-r,a+r)\times(r/8,7r/8)$ (recall (3.107)) we see that the set on the left-hand side is simply

$$[\tau(a-r)+A,\tau(a+r)+A]\times[\varepsilon r/2-7\tau r/8,\varepsilon r/2+7\tau r/8],$$

where we recalled that $z=(A,\varepsilon r/2,0)$ (see (3.111)). The second of these intervals is contained in $[\varepsilon r/2-\tau r,\varepsilon r/2+\tau r]=[0.02\varepsilon r,0.98\varepsilon r]$, where we recalled that $\tau=0.48\varepsilon$ (see (3.111)). Thus, (3.122) follows if the first of the intervals is contained in $[-(d-r),d-r]$, that is, if

$$|\tau a+A|\leq d-r-\tau r.$$

This last inequality follows from the fourth inequality in (3.110) and the facts that $\kappa\geq 10^4$ (recall (3.109)) and $\tau<1$, by writing

$$|\tau a+A|\leq\tau|a|+|A|=0.48\kappa r+|A|\leq 0.48\kappa r+0.05r<\kappa r/2<(\kappa-2)r=d-2r<d-r-\tau r.$$

The inclusions (3.114) and (3.122) combine to give

$$R^{-1}(\Gamma(G))\subset RECT\subset\overline{U_2},$$

and thus

$$\Gamma(G)\subset R(\overline{U_2})\subset G$$

(recall $G=R(\overline{U_1\cup U_2})$ and $\Gamma(x)=\tau x+z$), as required by the geometric arrangement, which is sketched in Fig. 3.14.

3.4.5 Construction of f_2, ϕ_2, T and Conclusion of the Arrangement

It remains to construct f_2, ϕ_2, T such that (v_2,f_2,ϕ_2) is a structure on U_2 and properties (3.51), (3.52) hold, that is,

$$f_2^2+Tv_2\cdot F^*>|v_2|^2\qquad\text{in }U_2, \tag{3.123}$$

and

$$f_2^2(y)+Tv_2(y)\cdot F^*(y)>\tau^{-2}\left(f_1(R^{-1}x)+f_2(R^{-1}x)\right)^2 \tag{3.124}$$

for

$$y = R^{-1}(\tau x + z), \quad x \in R\left(\overline{U_1} \cup \overline{U_2}\right),$$

respectively.

To this end, note that since U_2 is a rectangular ring, we can (as in Sect. 3.2.5) use Theorem 3.5 to obtain $f_2 \in C_0^\infty(P; [0, 1])$ such that supp $f_2 = \overline{U_2}$, $f_2 > 0$ in U_2, $f_2 = \mu$ on supp v_2, $Lf_2 > 0$ at points of U_2 of sufficiently small distance to ∂U_2. Here we choose $\mu > 100$ sufficiently large such that

$$\mu \geq 100\|f_1\|_\infty. \tag{3.125}$$

Following the recipe for a structure (Sect. 3.2.5), we let $\phi_2 \in C_0^\infty(U_2; [0, 1])$ be a cut-off function such that $\phi_2 = 1$ in supp v_2 and $Lf_2 > 0$ in $U_2 \setminus \{\phi_2 = 1\}$. Thus, (v_2, f_2, ϕ_2) is a structure on U_2. We now let

$$T := \frac{\mu^2 - 5}{1.1\varepsilon^2 B} \geq \frac{0.9\mu^2}{1.1\varepsilon^2 B} \tag{3.126}$$

and we verify (3.123) and (3.124).

Using (3.119) and the fact that $|v_2| \leq 2$ (recall Lemma 3.11 (iii)) we immediately obtain (3.123) by writing

$$f_2^2 + T v_2 \cdot F^* \geq \mu^2 - 1.1\varepsilon^2 BT = 5 > |v_2|^2 \quad \text{in supp } v_2,$$

and the claim in $U_2 \setminus \text{supp } v_2$ follows trivially from positivity of f_2 in U_2.

As for (3.124), we need to show

$$f_2^2(y) + T v_2(y) \cdot F^*(y) > \tau^{-2}\left(f_1(R^{-1}x) + f_2(R^{-1}x)\right)^2$$

for

$$y = R^{-1}(\tau x + z), \quad x \in R\left(\overline{U_1} \cup \overline{U_2}\right).$$

To this end, fix $x \in R\left(\overline{U_1} \cup \overline{U_2}\right)$. Since

$$\overline{U_1} \cup \overline{U_2} = \left(\overline{U_2} \cup \overline{U} \cup \overline{U^{a',r'}} \cup \overline{U^{a'',r''}}\right) \cup \overline{U^{a,r}}$$

we consider two cases.

Case 1. $x \in R\left(\overline{U_2} \cup \overline{U} \cup \overline{U^{a',r'}} \cup \overline{U^{a'',r''}}\right)$.

Then $R^{-1}x \in BOX$ and hence $y \in SBOX$ by (3.114). Thus, $v_2(y) = (1, 0)$ and $F_1^*(y) \geq 8B$ in $SBOX$ (see Lemma 3.11 (iv) and (3.118)) and using (3.126) and (3.125),

$$f_2^2(y) + T v_2(y) \cdot F^*(y) \geq 8TB \geq \frac{7.2}{1.1}\left(\frac{\mu}{\varepsilon}\right)^2 > \left(\frac{1.01}{0.48}\right)^2\left(\frac{\mu}{\varepsilon}\right)^2$$

$$= \tau^{-2}(1.01\mu)^2 \geq \tau^{-2}\left(\|f_2\|_\infty + \|f_1\|_\infty\right)^2.$$

Case 2. $x \in R\left(\overline{U^{a,r}}\right)$. Then $f_2(R^{-1}x) = 0$ (since $R^{-1}x \in \overline{U^{a,r}} \subset \overline{U_1}$ and $\overline{U_1}, \overline{U_2}$ are disjoint) and $y \in RECT$ (see (3.122)). Therefore, $v_2(y) = (1, 0)$, $F_1^*(y) \geq 0.01B$ (by Lemma 3.11 (iv) and (3.118)) and so using (3.126) and (3.125),

$$f_2^2(y) + T v_2(y) \cdot F^*(y) \geq 0.01 T B \geq \frac{0.009}{1.1} \left(\frac{\mu}{\varepsilon}\right)^2 > \left(\frac{0.01}{0.48}\right)^2 \left(\frac{\mu}{\varepsilon}\right)^2$$

$$= \tau^{-2}(0.01\mu)^2 \geq \tau^{-2} \|f_1\|_\infty^2 \geq \tau^{-2} \left(f_1(R^{-1}x) + f_2(R^{-1}x)\right)^2.$$

Hence, we obtain (3.124). This concludes the construction of geometric arrangement, and so also the proof of Theorem 3.1.

3.5   The Norm Inflation Result, Theorem 1.11

Here we prove Theorem 1.11, which is a corollary of Theorem 3.1.

Namely, given $\mathcal{N} > 0$ and $\vartheta > 0$ we will construct $T > 0$, $\nu_0 > 0$, $\eta > 0$ and divergence-free $u \in C^\infty(\mathbb{R}^3 \times (-\eta, T + \eta); \mathbb{R}^3)$ such that

$$-\frac{2\vartheta}{T^2 \nu_0} \leq \partial_t |u|^2 - 2\nu u \cdot \Delta u + u \cdot \nabla(|u|^2 + 2p) \leq 0 \qquad \text{in } \mathbb{R}^3 \times [0, T] \tag{3.127}$$

for all $\nu \in [0, \nu_0]$, $\operatorname{supp} u(t) \subset G$ for all $t \in [0, T]$ (where $G \subset \mathbb{R}^3$ is compact), and

$$\|u(T)\|_{L^\infty} \geq \mathcal{N} \|u(0)\|_{L^\infty} \tag{3.128}$$

Theorem 1.11 (which corresponds to the case $T = \nu_0 = 1$) then follows by a simple rescaling, namely,

$$\mathfrak{u}(x, t) := \sqrt{\nu_0 T} u \left(\sqrt{T/\nu_0} x, T t\right)$$

satisfies the claim of Theorem 1.11.

Let $T > 0$, $\nu_0 \in (0, 1)$, $\eta > 0$ and u be as in the proof of Theorem 3.1 (that is recall Proposition 3.8, (3.71), (3.126)). Note that this already gives that $\operatorname{div} u(t) = 0$, $\operatorname{supp} u(t) \subset G$ for all $t \in [0, T]$, smoothness of u and the rightmost inequality in (3.127). We now verify that such a choice satisfies the remaining two claims (i.e. the leftmost inequality in (3.127) and (3.128)) given we take ε (the "sharpness" of the geometric arrangement, recall (3.110)) and δ (a part of the definition of h, recall Lemma 3.7) sufficiently small.

Since

$$\|u(0)\|_\infty = \|h_0\|_\infty = \|f_1 + f_2\|_\infty,$$

and, from Proposition 3.8 (ii) and (3.67),

$$\|u(T)\|_\infty^2 \geq \|h_T\|_\infty^2 - \theta \geq \tau^{-2} \|f_1 + f_2\|_\infty^2,$$

we obtain

$$\|u(T)\|_\infty \geq \mathcal{N}\|u(0)\|_\infty$$

given $\tau^{-1} \geq \mathcal{N}$, that is, provided $\varepsilon > 0$ is sufficiently small such that

$$\varepsilon \leq (0.48\mathcal{N})^{-1},$$

in addition to the smallness requirements of the geometric arrangement (3.110). Note that making the value of ε smaller we also make T larger.

In order to obtain the leftmost inequality in (3.127), we perform similar calculation as in cases 1 and 2 in Sect. 3.3.2 given $\delta > 0$ is small as in Lemma 3.7 and additionally

$$\delta < \vartheta/2T^2.$$

Indeed, since (3.83) gives

$$2\nu_0 |u(x, 0, t) \cdot \Delta u(x, 0, t)| \leq \delta$$

and since $\nu_0 < 1$ (recall (3.71)) we write in the case $\phi_1(x) + \phi_2(x) < 1$ (that is Case 1 in Sect. 3.3.2)

$$
\begin{aligned}
\partial_t |u(x, 0, t)|^2 &= \partial_t q_{1,t}^k(x)^2 + \partial_t q_{2,t}^k(x)^2 \\
&= -2\delta(\phi_1(x) + \phi_2(x)) \\
&> -2\delta \\
&\geq -2\vartheta/T^2\nu_0 - u(x, 0, t) \cdot \nabla\left(|u(x, 0, t)|^2 + 2p(x, 0, t)\right) + 2\nu\, u(x, 0, t) \cdot \Delta u(x, 0, t)
\end{aligned}
$$

for all $\nu \in [0, \nu_0]$, $t \in [0, T]$, where we also used (3.44). In the case $\phi_1(x) + \phi_2(x) = 1$ (that is Case 2 in Sect. 3.3.2), we use (3.82) (in the same way as before) to obtain

$$
\begin{aligned}
\partial_t |u(x, 0, t)|^2 &= \partial_t q_{1,t}^k(x)^2 + \partial_t q_{2,t}^k(x)^2 \\
&= -2\delta - \left(a_1^k(t)v_1(x) + a_2^k(t)v_2(x)\right) \cdot \nabla\Big(h_{1,t}(x)^2 + h_{2,t}(x)^2 \\
&\qquad\qquad + 2p[a_1^k(t)v_1, h_{1,t}](x) + 2p[a_2^k(t)v_2, h_{2,t}](x)\Big) \\
&\geq -3\delta - \left(a_1^k(t)v_1(x) + a_2^k(t)v_2(x)\right) \cdot \nabla\Big(q_{1,t}^k(x)^2 + q_{2,t}^k(x)^2 \\
&\qquad\qquad + 2p[a_1^k(t)v_1, q_{1,t}^k](x) + 2p[a_2^k(t)v_2, q_{2,t}^k](x)\Big) \\
&= -3\delta - u(x, 0, t) \cdot \nabla\Big(|u(x, 0, t)|^2 + 2p(x, 0, t)\Big) \\
&\geq -2\vartheta/T^2\nu_0 - u(x, 0, t) \cdot \nabla\big(|u(x, 0, t)|^2 + 2p(x, 0, t)\big) + 2\nu\, u(x, 0, t) \cdot \Delta u(x, 0, t)
\end{aligned}
$$

for all $\nu \in [0, \nu_0]$, $t \in [0, T]$, where (as before) we also used (3.31) and (3.36) in the fourth step.

3.6 Appendix

The Function f Supported in $\overline{U}$ and with $Lf > 0$ Near ∂U

Here we show that for any set $U \Subset P$ of the shape of a rectangle or a "rectangular ring", that is, $U = V \setminus \overline{W}$ for some open rectangles V, W with $W \Subset V$, and any $\eta > 0$ there exists $\delta \in (0, \eta)$ and $f \in C_0^\infty(P; [0, 1])$ such that

$$\text{supp } f = \overline{U}, \quad f > 0 \text{ in } U \text{ with } f = 1 \text{ on } U_\eta$$

and

$$Lf > 0 \quad \text{in } U \setminus U_\delta.$$

(Recall that U_η denotes the η-subset of U, see (3.47))

The claim follows from Lemma 3.15 (which corresponds to the case of a rectangle) and from Lemma 3.16 (which corresponds to the case of a rectangle ring).

We will need a certain generalisation of the Mean Value Theorem. For $f : \mathbb{R} \to \mathbb{R}$, let $f[a, b]$ denote the finite difference of f on $[a, b]$,

$$f[a, b] := \frac{f(a) - f(b)}{a - b}$$

and let $f[a, b, c]$ denote the finite difference of $f[\cdot, b]$ on $[a, c]$,

$$f[a, b, c] := \left(\frac{f(a) - f(b)}{a - b} - \frac{f(c) - f(b)}{c - b} \right) \Big/ (a - c).$$

Lemma 3.13 (Generalised mean value theorem) *If $a < b < c$, f is continuous in $[a, c]$ and twice differentiable in (a, c) then there exists $\xi \in (a, c)$ such that $f[a, b, c] = f''(\xi)/2$.*

Proof We follow the argument of Theorem 4.2 in Conte and de Boor (1972). Let

$$p(x) := f[a, b, c](x - b)(x - c) + f[b, c](x - c) + f(c).$$

Then p is a quadratic polynomial approximating f at a, b, c, that is, $p(a) = f(a)$, $p(b) = f(b)$, $p(c) = f(c)$. Thus, the error function $e(x) := f(x) - p(x)$ has at least 3 zeros in $[a, c]$. A repeated application of Rolle's theorem gives that e'' has at least one zero in (a, c). In other words, there exists $\xi \in (a, c)$ such that $f''(\xi) = p''(\xi) = 2f[a, b, c]$. $\qquad\square$

Corollary 3.14 *If $f \in C^3$ is such that $f = 0$ on $(a - \delta, a]$ and $f''' > 0$ on $(a, a + \delta)$ for some $a \in \mathbb{R}$, $\delta > 0$ then*

$$\begin{cases} f''(x) > 0, \\ 0 < f'(x) < (x-a)f''(x), \qquad \text{for } x \in (a, a+\delta). \\ f(x) < (x-a)^2 f''(x) \end{cases}$$

Similarly, if $g = 0$ on $[a, a+\delta)$ and $g''' < 0$ on $(a-\delta, a)$ then

$$\begin{cases} g''(x) > 0, \\ 0 > g'(x) > (x-a)g''(x), \qquad \text{for } x \in (a-\delta, a). \\ g(x) < (x-a)^2 g''(x) \end{cases}$$

Proof Since $f''' > 0$ on $(a, a+\delta)$ we see that f'' is positive and increasing on this interval and so the first two claims for f follow from the mean value theorem. The last claim follows from the lemma above by noting that $2a - x \in (a-\delta, a]$ (so that $f(2a-x) = f(a) = 0$),

$$\begin{aligned} f(x) &= f(2a-x) - 2f(a) + f(x) \\ &= 2(x-a)^2 f[2a-x, a, x] \\ &= (x-a)^2 f''(\xi) \\ &< (x-a)^2 f''(x), \end{aligned}$$

where $\xi \in (2a-x, x)$. The claim for g follows by considering $f(x) := g(2a-x)$. $\qquad\square$

We now show the claim in the case of U in the shape of a rectangle.

Lemma 3.15 (The cut-off function on a rectangle) *Let $U \Subset P$ be an open rectangle, that is, $U = (a_1, b_1) \times (a_2, b_2)$ for some $a_1, a_2, b_1, b_2 \in \mathbb{R}$ with $b_1 > a_1, b_2 > a_2 > 0$. Given $\eta > 0$ there exists $\delta \in (0, \eta)$ and $f \in C_0^\infty(P; [0, 1])$ such that*

$$\operatorname{supp} f = \overline{U}, \quad f > 0 \text{ in } U \text{ with } f = 1 \text{ on } U_\eta,$$

$$Lf > 0 \quad \text{in } U \setminus U_\delta,$$

and f is symmetric with respect to the vertical axis of U, that is,

$$f\left(\frac{a_1+b_1}{2} - x_1, x_2\right) = f\left(\frac{a_1+b_1}{2} + x_1, x_2\right), \quad (x_1, x_2) \in P.$$

Proof Let $f_1, f_2 \in C_0^\infty(\mathbb{R}; [0, 1])$ be such that $\operatorname{supp} f_i = [a_i, b_i]$, $f_i > 0$ on (a_i, b_i) with $f_i = 1$ on $[a_i + \eta, b_i - \eta]$,

$$f_i''' > 0 \text{ on } (a_i, a_i + \varepsilon) \quad \text{and} \quad f_i''' < 0 \text{ on } (b_i - \varepsilon, b_i), \quad i = 1, 2,$$

for some $\varepsilon \in (0, \eta)$. (Take, for instance, f_i's such that

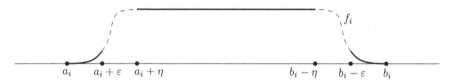

Fig. 3.15 The f_i's, $i = 1, 2$

$$f_i(x) = \begin{cases} 0 & x \le a_i, \\ \exp\left(-(x - a_i)^{-2}\right) & x \in (a_i, a_i + \varepsilon), \\ 1 & x \in (a_i + \eta, b_i - \eta), \\ \exp\left(-(b_i - x)^{-2}\right) & x \in (b_i - \varepsilon, b_i), \\ 0 & x \ge b_i, \end{cases}$$

where $\varepsilon \in (0, \eta)$ is sufficiently small such that $f_i \le 1$ on each of the intervals above, and define f_i on the remaining intervals $[a_i + \varepsilon, a_i + \eta]$, $[b_i - \eta, b_i - \varepsilon]$ in the way such that $f_i \in C^\infty$, $f_i \le 1$ and

$$f_i\left(\frac{a_i + b_i}{2} - x\right) = f_i\left(\frac{a_i + b_i}{2} + x\right), \quad x \in \mathbb{R},$$

$i = 1, 2$, see Fig. 3.15.

Let $f(x_1, x_2) := f_1(x_1) f_2(x_2)$. Clearly, supp $f = \overline{U}$, $f > 0$ in U, $f = 1$ on U_η and the last requirement of the lemma is satisfied due to the equality above. It remains to show that $Lf > 0$ on $U \setminus U_\delta$ for some $\delta > 0$. Let

$$g_1(x_1) := f_1''(x_1),$$
$$g_2(x_2) := f_2''(x_2) + f_2'(x_2)/x_2 - f_2(x_2)/x_2^2.$$

Then

$$Lf(x_1, x_2) = f_1''(x_1) f_2(x_2) + f_1(x_1) f_2''(x_2) + f_1(x_1) f_2'(x_2)/x_2 - f_1(x_1) f_2(x_2)/x_2^2$$
$$= g_1(x_1) f_2(x_2) + f_1(x_1) g_2(x_2).$$

Claim: There exists $d > 0$ such that

$$g_2 > f_2''/4 > 0 \qquad \text{on } (a_2, a_2 + d) \cup (b_2 - d, b_2).$$

The claim follows from the corollary of the generalised Mean Value Theorem (see Corollary 3.14 above) by writing, for $d > 0$ small such that $d < a_2/2$, $d < \varepsilon$ and $d/(b_2 - d) < 1/2$,

Fig. 3.16 The "d-corners" and "δ-stripes"

$$g_2(x_2) > f_2''(x_2) - f_2(x_2)/x_2^2$$

$$> f_2''(x_2)\left(1 - \left(\frac{x_2 - a_2}{x_2}\right)^2\right)$$

$$> f_2''(x_2)\left(1 - \left(\frac{d}{a_2}\right)^2\right)$$

$$> \frac{3}{4}f_2''(x_2)$$

$$> \frac{1}{4}f_2''(x_2)$$

$$> 0$$

for $x_2 \in (a_2, a_2 + d)$, and

$$g_2(x_2) = f_2''(x_2) + f_2'(x_2)/x_2 - f_2(x_2)/x_2^2 > f_2''(x_2)\left(1 + \frac{x_2 - b_2}{x_2} - \left(\frac{x_2 - b_2}{x_2}\right)^2\right)$$

$$> f_2''(x_2)\left(1 - \frac{d}{b_2 - d} - \left(\frac{d}{b_2 - d}\right)^2\right) > f_2''(x_2)/4 > 0$$

for $x_2 \in (b_2 - d, b_2)$.

Using the Claim and Corollary (3.14) (particularly positiveness of second derivatives) we see that g_i, f_i are positive on $(a_i, a_i + d) \cup (b_i - d, b_i)$, $i = 1, 2$. Thus,

$$Lf > 0 \quad \text{in } ((a_1, a_1 + d) \cup (b_1 - d, b_1)) \times ((a_2, a_2 + d) \cup (b_2 - d, b_2)),$$

that is, in the "d-corners" of U, see Fig. 3.16.

Now let $m, M > 0$ be small such that $f_i > m$, $|g_i| < M$ in $[a_i + d, b_i - d]$, $i = 1, 2$. Let $\delta \in (0, d)$ be such that $m/4 - \delta^2 M > 0$. The proof of the lemma is complete when we show that

$$Lf > 0 \quad \text{in } [a_i + d, b_i - d] \times ((a_j, a_j + \delta) \cup (b_j - \delta, b_j)), \ (i, j) = (1, 2), (2, 1),$$

that is, in the "δ-strips" at ∂U between the d-corners, see Fig. 3.16.

Let $x_1 \in [a_1 + d, b_1 - d]$ and $x_2 \in (a_2, a_2 + \delta)$. Then $g_1(x_1) > -M$, $g_2(x_2) > f_2''(x_2)/4$ (from *Claim*), $f_2(x_2) < (x_2 - a_2)^2 f_2''(x_2)$ (from the generalised Mean Value Theorem, see Corollary 3.14), $f_1(x_1) > m$, and so

$$
\begin{aligned}
Lf(x_1, x_2) &= g_1(x_1)f_2(x_2) + f_1(x_1)g_2(x_2) \\
&> -Mf_2(x_2) + f_1(x_1)f_2''(x_2)/4 \\
&> f_2''(x_2)\left(-M(x_2 - a_2)^2 + m/4\right) \\
&> f_2''(x_2)\left(m/4 - M\delta^2\right) \\
&> 0.
\end{aligned}
$$

As for $x_2 \in (b_2 - \delta, b_2)$, simply replace a_2 in the above calculation by b_2. The opposite case, that is, the case $x_1 \in (a_1, a_1 + \delta) \cup (b_1 - \delta, b_1)$, $x_2 \in [a_2 + d, b_2 - d]$, follows similarly. □

Let

$$
U^\eta := \{x \in \mathbb{R}^2 : \operatorname{dist}(x, U) < \delta\}
$$

denote the η-*neighbourhood* of U (this is not to confused with U_η which denotes the η-subset, recall the beginning of this section). We now extend the above lemma to the case of U in the shape of a "rectangular ring".

Lemma 3.16 (The cut-off function on a rectangular ring) *If $U \Subset P$ is a rectangular ring, that is $U = V \setminus \overline{W}$ where V, W are open rectangles with $W \Subset V$, then the assertion of the last lemma is valid.*

Proof It is enough to show that there exist $\delta > 0$ and $f \in C^\infty(P; [0, 1])$ such that $f = 0$ on $\overline{W}$, $f > 0$ outside $\overline{W}$ with $f = 1$ outside W^η and

$$
Lf > 0 \quad \text{in } W^\delta \setminus \overline{W}.
$$

Then the lemma follows by letting

$$
g := \begin{cases} \widetilde{f} & \text{on } P \setminus W^\eta, \\ f & \text{on } W^\eta, \end{cases}
$$

where $\widetilde{f}$ is from the previous lemma applied to V.

We write $W = (a_1, b_1) \times (a_2, b_2)$ for some $a_1, a_2, b_1, b_2 \in \mathbb{R}$ with $b_1 > a_1$ and $b_2 > a_2 > 0$. Let $f_1, f_2 \in C^\infty(\mathbb{R}; [0, 1])$ be such that $f_i = 1$ outside $(a_i - \eta, b_i + \eta)$, $f_i = 0$ on (a_i, b_i) and

$$
f_i''' < 0 \text{ on } (a_i - \varepsilon, a_i) \quad \text{on} \quad f_i''' > 0 \text{ on } (b_i, b_i + \varepsilon), \quad i = 1, 2,
$$

for some $\varepsilon \in (0, \eta/2)$. (Such functions can be constructed by use of the exponential function, as in the previous lemma, see also Fig. 3.17.) Let $f(x_1, x_2) := f_1(x_1)f_2(x_2)$.

Fig. 3.17 The f_i's, $i = 1, 2$ (cf. Fig. 3.15)

Then $f = 0$ on $\overline{W}$, $f > 0$ outside $\overline{W}$ with $f = 1$ outside W^η. It remains to show that $Lf > 0$ in $W^\delta \setminus \overline{W}$ for some $\delta > 0$. Note that

$$Lf(x_1, x_2) = \left(f_1''(x_1) - f_1(x_1)/x_2^2\right) + \left(f_2''(x_2) + f_2'(x_2)/x_2 - f_2(x_2)/x_2^2\right)$$
$$=: g_1(x_1, x_2) + g_2(x_2).$$

As in *Claim* in the proof of the previous lemma, we see that

$$g_2 > f_2''/4 > 0 \quad \text{in } (a_2 - \delta, a_2) \cup (b_2, b_2 + \delta)$$

for sufficiently small $\delta > 0$. Thus, since f_2 vanishes on $[a_2, b_2]$, we see that

$$g_2 \geq 0 \text{ on } (a_2 - \delta, b_2 + \delta) \quad \text{with} \quad g_2 > 0 \text{ outside } [a_2, b_2]. \tag{3.129}$$

As for g_1 let δ be such that $\delta/(a_2 - \delta) < 1/2$. Then, using the corollary of the generalised Mean Value Theorem (Corollary 3.14) we obtain for any $x_2 > a_2 - \delta$

$$g_1(x_1, x_2) = f_1''(x_1) - f_1(x_1)/x_2^2 > f_1''(x_1)\left(1 - \left(\frac{x_1 - a_1}{x_2}\right)^2\right)$$

$$> f_1''(x_1)\left(1 - \left(\frac{\delta}{a_2 - \delta}\right)^2\right) > \frac{3}{4}f_1''(x_1) > 0$$

for $x_1 \in (a_1 - \delta, a_1)$. As for $x_1 \in (b_1, b_1 + \delta)$ replace a_1 in the above calculation by b_1. Thus, since f_1 vanishes on $[a_1, b_1]$ we see that for each $x_2 > a_2 - \delta$

$$g_1(\cdot, x_2) \geq 0 \text{ on } (a_1 - \delta, b_1 + \delta) \quad \text{with} \quad g_1(\cdot, x_2) > 0 \text{ outside } [a_1, b_1].$$

This and (3.129) give

$$Lf \geq 0 \text{ on } W^\delta \quad \text{with} \quad Lf > 0 \text{ outside } \overline{W},$$

as required.

$\square$

Preliminary Calculations

Here we briefly verify (3.19). Let $\varphi \in [0, 2\pi)$ and let $R := R_\varphi$ for brevity of notation. We can represent R in the matrix form

$$R = \begin{pmatrix} 1 & 0 & 0 \\ 0 & \cos\varphi & -\sin\varphi \\ 0 & \sin\varphi & \cos\varphi \end{pmatrix}.$$

Note that R is orthogonal, so that $R^T R = I$, where I denotes the identity matrix. We write

$$\nabla u := \begin{pmatrix} \partial_1 u_1 & \partial_2 u_1 & \partial_3 u_1 \\ \partial_1 u_2 & \partial_2 u_2 & \partial_3 u_2 \\ \partial_1 u_3 & \partial_2 u_3 & \partial_3 u_3 \end{pmatrix}, \qquad \nabla q := \begin{pmatrix} \partial_1 q \\ \partial_2 q \\ \partial_3 q \end{pmatrix};$$

If $u(Rx) = Ru(x)$ and $q(Rx) = q(x)$ we can use the calculus identities

$$\nabla(u(Rx)) = \nabla u(Rx)R, \qquad \nabla(Ru) = R\nabla u \qquad \text{and} \qquad \nabla(q(Rx)) = R^T \nabla q(Rx)$$

to write

$$((u \cdot \nabla)u)(Rx) = \nabla u(Rx)u(Rx) = \nabla(u(Rx))R^T u(Rx) = \nabla(Ru(x))R^T Ru(x)$$
$$= R\nabla u(x)u(x) = R((u \cdot \nabla)u)(x),$$

$$|u|^2(Rx) = u(Rx)^T u(Rx) = (Ru(x))^T(Ru(x)) = u(x)^T u(x) = |u|^2(x),$$

$$\text{div } u(Rx) = \text{tr}\nabla u(Rx) = \text{tr}(\nabla(u(Rx))R^T) = \text{tr}(\nabla(Ru(x))R^T) = \text{tr}(R\nabla u(x)R^T)$$
$$= \text{tr}\nabla u(x) = \text{div } u(x),$$

where we used the fact that $\text{tr}\, RAR^T = A$ for any matrix A,

$$(u \cdot \nabla q)(Rx) = u(Rx)^T \nabla q(Rx) = (Ru(x))^T(R\nabla(q(Rx))) = u(x)^T \cdot \nabla q(x) = (u \cdot \nabla)q.$$

By taking $q := |u|^2$ we obtain

$$(u \cdot \nabla |u|^2)(Rx) = (u \cdot \nabla |u|^2)(x).$$

Also, since $u(x) = R^T u(Rx)$, we have

$$\Delta u_k(x = \Delta(u_k(x)) = \sum_j \Delta(R_{jk}u_j(Rx)) = \sum_{i,j} \partial_i \partial_i (R_{jk}u_j(Rx))$$

$$= \sum_{i,j,l} R_{jk} R_{li} \partial_i (\partial_l u_j(Rx)) = \sum_{i,j,l,m} R_{jk} R_{li} R_{mi} \partial_m \partial_l u_j(Rx)$$

$$= \sum_{j,l,m} R_{jk} \delta_{ml} \partial_m \partial_l u_j(Rx) = \sum_j R_{jk} \Delta u_j(Rx)$$

for each $k \in \{1, 2, 3\}$, where δ_{ml} denotes the Kronecker delta. Thus

$$\Delta u(Rx) = R\Delta u(x),$$

as needed. Finally

$$(u \cdot \Delta u)(Rx) = u(Rx)^T \Delta u(Rx) = (Ru(x))^T R\Delta u(x) = u(x)^T \Delta u(x) = (u \cdot \Delta u)(x).$$

Some Continuity Properties of $p[v, f]$ and $u[v, f]$ with Respect to f

In this section, we discuss two rather technical results regarding continuity of $\nabla p[v, f]$ with respect to f (Lemma 3.17 below) as well as continuity of the term $u[v, f] \cdot \Delta u[v, f]$ with respect to f (Lemma 3.18). We mentioned them at the end of Sect. 3.2.3. They are used in (3.82), (4.35) and (3.83), (4.36), respectively.

Lemma 3.17 *Suppose that $v_k \in C^\infty(P; \mathbb{R}^2)$, $f_k, f \in C^\infty(P; [0, \infty))$ are such that $|v_k| < \min(f_k, f)$ and*

$$\operatorname{supp} v_k, \operatorname{supp} f_k, \operatorname{supp} f \subset K$$

for all k, where K is a compact subset of P, and that

$$f_k \to f, \quad \text{and} \quad \nabla f_k \to f \quad \text{uniformly in } P.$$

Then

$$\nabla p[v_k, f_k] - \nabla p[v_k, f] \to 0$$

uniformly on $\mathbb{R}^2$.

The point of the lemma is that we do not require any convergence of the v_k's. We used the lemma (or rather its straightforward generalisation to include time dependence) in a convergence argument in (3.82).

Proof Let $z = (z_1, z_2) \in \overline{P}$ and observe that for $u = u[v, f]$ (for some $v = (v_1, v_2)$, f) we have the identity

$$\sum_{i,j=1}^{3} \partial_i u_j(z_1, z_2, 0)\partial_j u_i(z_1, z_2, 0)$$

$$= (\partial_1 v_1)^2 + (\partial_2 v_2)^2 + 2\partial_2 v_1 \partial_1 v_2 - \left(\partial_2 f^2 - \partial_2 |v|^2\right)/2z_2 + v_2^2/z_2^2,$$

where (for brevity) we skipped the argument z of the functions on the right-hand side. Therefore, setting

$$F_k(y) := \sum_{i,j=1}^{3} \partial_i u_j[v_k, f_k](y)\partial_j u_i[v_k, f_k](y),$$

$$G_k(y) := \sum_{i,j=1}^{3} \partial_i u_j[v_k, f](y)\partial_j u_i[v_k, f](y),$$

where $y \in \mathbb{R}^3$, and letting $z = (z_1, z_2, 0) := R^{-1}(y)$ we obtain, using axisymmetry of F_k, G_k (recall (3.19))

$$F_k(y) - G_k(y) = F_k(z) - G_k(z)$$
$$= \left(\partial_2 f^2 - \partial_2 f_k^2\right)/2z_2,$$

as all the term that include the components of v_k's vanish. Thus

$$F_k - G_k \to 0 \qquad \text{uniformly in } \mathbb{R}^3.$$

Since

$$\nabla p^*[v_k, f_k](x) - \nabla p^*[v_k, f](x) = \frac{1}{4\pi}\int_{\mathbb{R}^3} (F_k(y) - G_k(y))\frac{x - y}{|x - y|^3}dy$$

we immediately obtain that

$$\nabla p^*[v_k, f_k] - \nabla p^*[v_k, f] \to 0 \qquad \text{uniformly on } \mathbb{R}^3,$$

and the claim of the lemma follows by setting $x_3 = 0$.   □

We now a result regarding continuity of $u[v, f] \cdot \Delta u[v, f]$ with respect to f.

Lemma 3.18 *Let $T > 0$, K be a compact subset of P and suppose that $v \in C^\infty(P; \mathbb{R}^2)$ and $f_k, f: K \times [0, T] \to [0, \infty)$ are such that $|v| < f(t)$ on K for every $t \in [0, T]$, and let (for each k) $a_k: [0, T] \to [-1, 1]$ be arbitrary. If*

$$D^\alpha f_k \to D^\alpha f, \text{ uniformly in } K \times [0, T], \tag{3.130}$$

for any multiindex $\alpha = (\alpha_1, \alpha_2)$ with $|\alpha| \leq 2$, then

$$u[a_k(t)v, f_k(t)] \cdot \Delta u[a_k(t)v, f_k(t)] - u[a_k(t)v, f(t)] \cdot \Delta u[a_k(t)v, f(t)] \to 0$$

uniformly on $R(K)$ and in $t \in [0, T]$.

Proof Since in cylindrical coordinates

$$\Delta = \partial_{x_1 x_1} + \partial_{\rho\rho} + \frac{1}{\rho}\partial_\rho + \frac{1}{\rho^2}\partial_{\varphi\varphi}$$

we obtain, using (3.40),

$$\Delta u_1[v, f](x_1, \rho, 0) = \left(\partial_{x_1 x_1} + \partial_{\rho\rho} + \rho^{-1}\partial_\rho\right) v_1,$$

$$\Delta u_2[v, f](x_1, \rho, 0) = \left(\partial_{x_1 x_1} + \partial_{\rho\rho} + \rho^{-1}\partial_\rho - \rho^{-2}\right) v_2,$$

$$\Delta u_3[v, f](x_1, \rho, 0) = \left(\partial_{x_1 x_1} + \partial_{\rho\rho} + \rho^{-1}\partial_\rho - \rho^{-2}\right)\sqrt{f^2 - |v|^2},$$

compare with (3.41). Thus

$$u[v, f] \cdot \Delta u[v, f] = (v\text{-terms}) + \sqrt{f^2 - |v|^2}\left(\partial_{x_1 x_1} + \partial_{\rho\rho} + \rho^{-1}\partial_\rho - \rho^{-2}\right)\sqrt{f^2 - |v|^2},$$

where we skipped the argument "$(x_1, \rho, 0)$" on the left-hand side and we denoted all terms involving components of v (and its derivatives) by "(v-terms)". Expanding the last term on the right-hand side we obtain

$$(\partial_{x_1 x_1} + \partial_{\rho\rho})f^2/2 - \frac{\left(\partial_{x_1}(f^2 - |v|^2)\right)^2 + \left(\partial_\rho(f^2 - |v|^2)\right)^2}{4(f^2 - |v|^2)} + \rho^{-1}f\partial_\rho f - \rho^{-2}f^2 + (v\text{-terms}).$$

Hence, since both

$$F_k(x, t) := u[a_k(t)v, f_k(t)](x) \cdot \Delta u[a_k(t)v, f_k(t)](x), \quad \text{and}$$

$$G_k(x, t) := u[a_k(t)v, f(t)](x) \cdot \Delta u[a_k(t)v, f(t)](x)$$

are axisymmetric (recall (3.19)) we can write

$$\begin{aligned}
F_k(x, t) - G_k(x, t) = &F_k(x_1, \rho, 0, t) - G_k(x_1, \rho, 0, t) \\
= &(\partial_{x_1 x_1} + \partial_{\rho\rho})(f_k^2 - f^2)/2 \\
&+ \frac{\left(\partial_{x_1}(f^2 - |a_k v|^2)\right)^2 + \left(\partial_\rho(f^2 - |a_k v|^2)\right)^2}{4(f^2 - |a_k v|^2)} \\
&- \frac{\left(\partial_{x_1}(f_k^2 - |a_k v|^2)\right)^2 + \left(\partial_\rho(f_k^2 - |a_k v|^2)\right)^2}{4(f_k^2 - |a_k v|^2)} \\
&+ \rho^{-1}(f_k\partial_\rho f_k - f\partial_\rho f) - \rho^{-2}(f_k^2 - f^2)
\end{aligned}$$

since the v-terms cancel out. Here $\rho := \sqrt{x_2^2 + x_3^2}$. Thus, the claim of the lemma (i.e. that $F_k - G_k \to 0$ uniformly in $R(K) \times [0, T]$) follows if we can show that the difference of the two fractions above converges uniformly to 0 (the other terms converge by assumption). We will focus on the terms with derivatives with respect to ρ, and we will write $\partial \equiv \partial_\rho$ for brevity (the terms with the x_1 derivatives are analogous). Bringing the two fractions under the common denominator we obtain

$$\frac{\left(\partial(f^2 - |a_k v|^2)\right)^2 (f_k^2 - |a_k v|^2) - \left(\partial(f_k^2 - |a_k v|^2)\right)^2 (f^2 - |a_k v|^2)}{4(f_k^2 - |a_k v|^2)(f^2 - |a_k v|^2)}.$$

The denominator is bounded away from 0 (as $|a_k| \leq 1$, $|v| < f$ on the compact set $K \times [0, T]$ and $f_k \to f$ uniformly), and so it is sufficient to verify that the numerator converges uniformly, which is clear after expanding the brackets the terms with v only cancel out and (since no derivative falls on a_k; recall a_k depends on time only) each of the other terms is of the form

(something converging uniformly by (3.30)) $*$ (term involving a_k and derivatives of v)

which converge uniformly to 0 (as $|a_k| \leq 1$), as required.   □

Chapter 4
Weak Solution of the Navier–Stokes Inequality with a Blow-Up on a Cantor Set

In this chapter, we construct Scheffer's counterexample (Theorem 1.8), that is, a weak solution of the Navier–Stokes inequality that blows up on a Cantor set $S \times \{T_0\}$ with $d_H(S) \geq \xi$ for any preassigned $\xi \in (0, 1)$. Such a solution was first constructed by Scheffer (1987), and we now state the result in a precise form.

Theorem 4.1 (Nearly one-dimensional singular set) *Given any $\xi \in (0, 1)$ there exists $\nu_0 > 0$, a compact set $G \Subset \mathbb{R}^3$ and a function $\mathfrak{u} \colon \mathbb{R}^3 \times [0, \infty) \to \mathbb{R}^3$ that is a weak solution of the Navier–Stokes inequality for any $\nu \in [0, \nu_0]$ such that $\mathfrak{u}(t) \in C^\infty$, $\operatorname{supp} \mathfrak{u}(t) \subset G$ for all t, and*

$$\xi \leq d_H(S) \leq 1,$$

where

$$S := \{(x, t) \in \mathbb{R}^3 \times (0, \infty) : \mathfrak{u}(x, t) \text{ is unbounded in any neighbourhood of } (x, t)\}.$$

Recall that the difference between this theorem and the result of the previous chapter (Theorem 3.1) is the size of the singular set. In the case of Theorem 3.1, the singular set is a point $\{x_0\} \times \{T_0\}$ and in the case of Theorem 4.1 it is a set $S \times \{T_0\}$, where S is a Cantor set with $d_H(S) \in [\xi, 1]$ for given $\xi \in (0, 1)$. We will show how Theorem 4.1 can be obtained by sharpening the proof of Theorem 3.1 (which we shall refer to by writing "previously") as intuitively sketched in Fig. 4.1.

In other words, the solution $\mathfrak{u}$ (of Theorem 4.1) is obtained by a similar switching procedure as in Sect. 3.1, except that at every switching the support of $\mathfrak{u}$ shrinks (by a fixed factor) to form M copies of itself ($M \in \mathbb{N}$) and thus form a Cantor set S at the limit $t \to T_0^-$. It is remarkable that such approach allows enough freedom to ensure that $d_H(S)$ is arbitrarily close to 1 (from below). Before proceeding to the proof, we

© Springer Nature Switzerland AG 2019
W. S. Ożański, *The Partial Regularity Theory of Caffarelli, Kohn, and Nirenberg and its Sharpness*, Advances in Mathematical Fluid Mechanics, https://doi.org/10.1007/978-3-030-26661-5_4

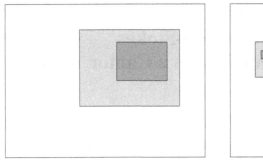
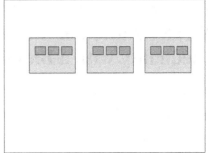

Fig. 4.1 The (intuitive) difference in constructing solutions to Theorem 3.1 (left; cf. Fig. 3.1) and Theorem 4.1 (right)

briefly comment on the construction of such a Cantor set and we introduce some handy notation. We then prove Theorem 4.1 in Sect. 4.2.

4.1 Constructing a Cantor Set

The problem of constructing Cantor sets is usually demonstrated in a one-dimensional setting using intervals, as in the following proposition.

Proposition 4.2 *Let $I \subset \mathbb{R}$ be an interval and let $\tau \in (0, 1)$, $M \in \mathbb{N}$ be such that $\tau M < 1$. Let $C_0 := I$ and consider the iteration in which in the j-th step ($j \geq 1$) the set C_j is obtained by replacing each interval J contained in the set C_{j-1} by M equidistant copies of τJ contained in J, see, for example, Fig. 4.2. Then the limiting object*

$$C := \bigcap_{j \geq 0} C_j$$

is a Cantor set whose Hausdorff dimension equals $- \log M / \log \tau$.

See Example 4.5 in Falconer (2014) for a proof. Thus if $\tau \in (0, 1)$, $M \in \mathbb{N}$ satisfy

$$\tau^{\xi} M \geq 1 \quad \text{for some } \xi \in (0, 1),$$

we obtain a Cantor set C with

$$d_H(C) \geq \xi. \tag{4.1}$$

Note that both the above inequality and the constraint $\tau M < 1$ (which is necessary for the iteration described in the proposition above, see also Fig. 4.2) can be satisfied only for $\xi < 1$. In the remainder of this section, we extend the result from the proposition above to the three-dimensional setting.

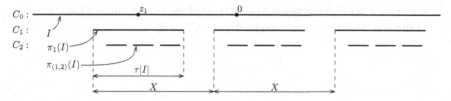

Fig. 4.2 A construction of a Cantor set C on a line (here $M = 3$, $j = 0, 1, 2$)

Let $G \subset \mathbb{R}^3$ be a compact set. We will later take $G := R(\overline{U}_1 \cup \overline{U}_2)$ (as in the case of Theorem 3.1), and so for convenience suppose further that $G = G_1 \cup G_2$ for some disjoint compact sets $G_1, G_2 \subset \mathbb{R}^3$, and such that $G_2 = R(\overline{U}_2)$ for some open and connected $U_2 \Subset P$. Let $\tau \in (0, 1)$, $M \in \mathbb{N}$, $z = (z_1, z_2, 0) \in G_2$, $X > 0$ be such that

$$\tau^\xi M \geq 1, \qquad \tau M < 1 \tag{4.2}$$

and

$$\{\Gamma_n(G)\}_{n=1,\ldots,M} \text{ is a family of pairwise disjoint subsets of } G_2, \tag{4.3}$$

where

$$\Gamma_n(x) := \tau x + z + (n-1)(X, 0, 0).$$

Equivalently,

$$\Gamma_n(x_1, x_2, x_3) = (\beta_n(x_1), \gamma(x_2), \tau x_3), \tag{4.4}$$

where

$$\begin{cases} \beta_n(x) := \tau x + z_1 + (n-1)X, \\ \gamma(x) := \tau x + z_2, \end{cases} \qquad x \in \mathbb{R}, n = 1, \ldots, M.$$

Now for $j \geq 1$ let

$$M(j) := \big\{ m = (m_1, \ldots, m_j) \colon m_1, \ldots, m_j \in \{1, \ldots, M\} \big\}$$

denote the set of multi-indices m. Note that in particular $M(1) = \{1, \ldots, M\}$. Casually speaking, each multiindex $m \in M(j)$ plays the role of a "coordinate" which let us identify any component of the set obtained in the j-th step of the construction of the Cantor set. Namely, letting

$$\pi_m := \beta_{m_1} \circ \ldots \circ \beta_{m_j}, \qquad m \in M(j),$$

that is

$$\pi_m(x) = \tau^j x + z_1 \frac{1 - \tau^j}{1 - \tau} + X \sum_{k=1}^{j} \tau^{k-1}(m_k - 1), \quad x \in \mathbb{R}, \tag{4.5}$$

we see that the set C_j obtained in the j-th step of the construction of the Cantor set C (from the proposition above) can be expressed simply as

$$C_j := \bigcup_{m \in M(j)} \pi_m(I),$$

see Fig. 4.2. Moreover, each $\pi_m(I)$ can be identified by, roughly speaking, first choosing m_1-th subinterval, then m_2-th subinterval, ... , up to m_j-th interval, where $m = (m_1, \dots, m_j)$. This is demonstrated in Fig. 4.2 in the case when $m = (1, 2) \in M(2)$.

In order to proceed with our construction of a Cantor set in three dimensions, let

$$\Gamma_m(x_1, x_2, x_3) := \left(\pi_m(x_1), \gamma^j(x_2), \tau^j x_3 \right).$$

Note that such a definition reduces to (4.4) in the case $j = 1$. If $j = 0$ then let $M(0)$ consist of only one element m_0 and let $\pi_{m_0} := \mathrm{id}$. Moreover, if $m \in M(j)$ and $\overline{m} \in M(j-1)$ is its sub-multiindex, that is, $\overline{m} = (m_1, \dots, m_{j-1})$ ($\overline{m} = m_0$ if $j = 1$), then (4.3) gives

$$\Gamma_m(G) = \Gamma_{\overline{m}}(\Gamma_{m_j}(G)) \subset \Gamma_{\overline{m}}(G_2), \tag{4.6}$$

which is a three-dimensional equivalent of the relation $\pi_m(I) \subset \pi_{\overline{m}}(I)$ (see Fig. 4.2). The above inclusion and (4.3) give that

$$\Gamma_m(G) \cap \Gamma_{\widetilde{m}}(G) = \emptyset \qquad \text{for } m, \widetilde{m} \in M(j), \ j \geq 1, \text{ with } m \neq \widetilde{m}. \tag{4.7}$$

Another consequence of (4.6) is that the family of sets

$$\left\{ \bigcup_{m \in M(j)} \Gamma_m(G) \right\}_j \qquad \text{decreases as } j \text{ increases.} \tag{4.8}$$

Moreover, given j, each of the sets $\Gamma_m(G)$, $m \in M(j)$, is separated from the rest by at least $\tau^{j-1}\zeta$, where $\zeta > 0$ is the distance between $\Gamma_n(G)$ and $\Gamma_{n+1}(G)$, $n = 1, \dots, M-1$ (recall (4.3)).

Taking the intersection in j, we obtain

$$S := \bigcap_{j \geq 0} \bigcup_{m \in M(j)} \Gamma_m(G), \tag{4.9}$$

and we now show that

$$\xi \leq d_H(S) \leq 1. \tag{4.10}$$

Noting that S is a subset of a line, the upper bound is trivial. As for the lower bound note that

$$S \supset \bigcap_{j \geq 0} \bigcup_{m \in M(j)} \Gamma_m(G_2) =: S'.$$

Thus, letting $I \subset \mathbb{R}$ be the orthogonal projection of G_2 onto the x_1 axis, we see that I is an interval (since U_2 is connected). Thus the orthogonal projection of S' onto the x_1 axis is

$$\bigcap_{j \geq 0} \bigcup_{m \in M(j)} \pi_m(I) = C,$$

where C is as in the proposition above. Thus, since the orthogonal projection onto the x_1 axis is a Lipschitz map, we obtain $d_H(S') \geq d_H(C)$ (as a property of Hausdorff dimension, see, for example, Proposition 3.3 in Falconer 2014). Consequently

$$d_H(S) \geq d_H(S') \geq d_H(C) \geq \xi,$$

as required (recall (4.1) for the last inequality).

4.2 Sketch of the Proof of Theorem 4.1

As in the proof of Theorem 3.1, the proof is based on a geometric arrangement. Here we will need a certain sharper geometric arrangement as follows.

> By the *geometric arrangement* (for Theorem 4.1), we mean a pair open sets $U_1, U_2 \Subset P$ together with the corresponding structures (v_1, f_1, ϕ_1), (v_2, f_2, ϕ_2) such that $\overline{U_1} \cap \overline{U_2} = \emptyset$ and, for some $T > 0$, $X > 0$, $\tau \in (0, 1)$, $z = (z_1, z_2, 0) \in \mathbb{R}^3$, $M \in \mathbb{N}$, (4.2) and (4.3) hold with
>
> $$G = G_1 \cup G_2 := R\left(\overline{U_1} \cup \overline{U_2}\right),$$
>
> $$f_2^2 + T v_2 \cdot F[v_1, f_1] > |v_2|^2 \quad \text{in } U_2 \qquad (4.11)$$
>
> and
>
> $$f_2^2(y_n) + T v_2(y_n) \cdot F[v_1, f_1](y_n) > \tau^{-2}\left(f_1(R^{-1}x) + f_2(R^{-1}x)\right)^2 \quad (4.12)$$
>
> for all $x \in G$ and $n = 1, \ldots, M$, where
>
> $$y_n = R^{-1}(\Gamma_n(x)) \qquad (4.13)$$
>
> and Γ_n is as in (4.4).

The difference, as compared to the previous geometric arrangement (see Sect. 3.3), is (4.3) and (4.12), which we now require for all $n = 1, \ldots, M$ (rather than only for $n = 1$, which was the case previously). This reflects the fact that at each switching time we expect the support of u to form M copies of itself (rather than one copy, which was the case in Theorem 3.1). Except for this, the inequalities (4.2) specify the relation between τ and M that needs to be satisfied in order to obtain blow-up on a Cantor set with Hausdorff dimension at least ξ. In fact, the previous geometric arrangement is recovered if one takes $\xi = 0$, $M = 1$. We now show how Theorem 4.1 follows (given the geometric arrangement) in a similar way as discussed in Sect. 3.3, except for a subtle change in the construction of the vector field $u^{(j)}$ (recall the previous construction (3.8)).

To this end, as in Sect. 3.3, let $\theta > 0$ be sufficiently small such that

$$f_2^2(y_n) + T v_2(y_n) \cdot F[v_1, f_1](y_n) > \tau^{-2} \left(f_1(R^{-1}x) + f_2(R^{-1}x) \right)^2 + 2\theta \quad (4.14)$$

for $x \in G$, $n = 1, \ldots, M$ (by (4.12)), and set

$$h_t := h_{1,t} + h_{2,t}, \quad (4.15)$$

where h_1, h_2 are given by (3.64), (3.65), that is

$$
\begin{aligned}
h_{1,t}^2 &:= f_1^2 - 2t\delta\phi_1, \\
h_{2,t}^2 &:= f_2^2 - 2t\delta\phi_2 + \int_0^t v_2 \cdot F[v_1, h_{1,s}] \, ds.
\end{aligned}
\quad (4.16)
$$

As in Lemma 3.7, let $\delta > 0$ be sufficiently small so that $h_1, h_2 \in C^\infty(P \times (-\delta, T + \delta); [0, \infty))$,

$$(v_i, h_{i,t}, \phi_i) \text{ is a structure on } U_i \qquad \text{for } t \in (-\delta, T + \delta), i = 1, 2,$$

and

$$h_{2,T}^2(y_n) > \tau^{-2} \left(f_1(R^{-1}x) + f_2(R^{-1}x) \right)^2 + \theta \quad (4.17)$$

for $x \in G, n = 1, \ldots, M$. Here only the last inequality differs from the corresponding property (3.67); note, however, that this is a consequence of (4.14), as previously (3.67) was a consequence of (3.62).

Let $\nu_0 > 0$ be as in (3.71). As in Sect. 3.1, in order to obtain a solution u, we want to find $\eta_j > 0$ and a velocity field $u^{(j)}$. However, in contrast to the arguments from Sect. 3.1, the velocity field $u^{(j)}$ will not be obtained by rescaling a single vector field u (recall (3.8)), which we have pointed out above. In fact, for each j we expect $u^{(j)}$ to consist of M^j disjointly supported vector fields (recall the comments preceding

Sect. 4.1). A naive idea of constructing $u^{(j)}$ would be to consider M^j rescaled copies of u, that is the vector field

$$\tilde{u}^{(j)}(x, t) := \tau^{-j} \sum_{m \in M(j)} u(\Gamma_m^{-1}(x), \tau^{-2j}(t - t_j)), \qquad j \geq 0.$$

For such vector field

$$\operatorname{supp} \tilde{u}^{(j)}(t) = \bigcup_{m \in M(j)} \Gamma_m(G), \qquad t \in [t_j, t_{j+1}], j \geq 0,$$

which shrinks to the Cantor set S as $j \to \infty$ (recall (4.9)), as expected. However, the observation that the pressure function does not have a local character (that is, the pressure function corresponding to a compactly supported vector field does not have compact support, recall (3.2)) suggests that $\tilde{u}$ has little chance to satisfy the local energy inequality (3.1). Instead, one needs to make use of the following proposition.

Proposition 4.3 *Let $j \geq 0$ and*

$$h_t^{(j)}(x_1, x_2) := \sum_{m \in M(j)} h_t(\pi_m^{-1}(\tau^j x_1), x_2), \tag{4.18}$$

where h_t is given by (4.15), $t \in (-\delta, T + \delta)$. Then there exists a vector field $v^{(j)} \in C^\infty \left(\mathbb{R}^3 \times [0, T]; \mathbb{R}^3 \right)$ such that

(i) $\operatorname{div} v^{(j)}(t) = 0$ and $\operatorname{supp} v^{(j)}(t) = R \left(\operatorname{supp} h_t^{(j)} \right)$, $t \in [0, T]$,
(ii) for all $x \in \mathbb{R}^3$, $t \in [0, T]$

$$\left| v^{(j)}(x, 0) \right| = h_0^{(j)} \left(R^{-1} x \right), \qquad \left| \left| v^{(j)}(x, t) \right|^2 - h_t^{(j)} \left(R^{-1} x \right)^2 \right| < \theta,$$

(iii) the Navier–Stokes inequality

$$\partial_t \left| v^{(j)} \right|^2 \leq -v^{(j)} \cdot \nabla \left(\left| v^{(j)} \right|^2 + 2\overline{p}^{(j)} \right) + 2\nu \, v^{(j)} \cdot \Delta v^{(j)}$$

is satisfied in $\mathbb{R}^3 \times [0, T]$ for every $\nu \in [0, \nu_0]$, and
(iv) $\|v^{(j)}(t)\|_{L^2} \leq C$ for $t \in [0, T]$ and

$$\int_0^T \|\nabla v^{(j)}(t)\|_{L^2}^2 dt \leq C,$$

for some constant $C > 0$ which is independent of j, where $\overline{p}^{(j)}$ is the pressure function corresponding to $v^{(j)}$.

We prove the proposition in Sect. 4.3. Proposition 4.3 is a generalisation of the previous result (Proposition 3.8, which is recovered by taking $j = 0$). In the previous construction, $u^{(j)}$ (i.e. the solution for times between t_j and t_{j+1}, recall Fig. 3.1) was given by

$$u^{(j)}(x, t) := \tau^{-j} v^{(0)} \left(\Gamma^{-j}(x), \tau^{-2j}(t - t_j) \right), \qquad j \geq 1 \qquad (4.19)$$

(recall (3.8)), where $v^{(0)}$ is from the proposition above (or equivalently u from Proposition 3.8). Here given $j \geq 0$ we set

$$u^{(j)}(x_1, x_2, x_3, t) := \tau^{-j} v^{(j)}(\tau^{-j} x_1, \gamma^{-j}(x_2), \tau^{-j} x_3, \tau^{-2j}(t - t_j)) \qquad (4.20)$$

(where $t_0 := 0$ and $t_j := T \sum_{k=0}^{j-1} \tau^{2k}$, as previously), which (at each time t) consists of M^j disjointly supported (in rescaled copies of $G = R(\overline{U}_1 \cup \overline{U}_2)$) vector fields and we ensure (by constructing $v^{(j)}$ in Proposition 4.3) that such a "composite vector field" still satisfies the Navier–Stokes inequality despite the non-local character of the pressure function. Indeed (as in Sect. 3.3 (cf. Proposition 3.8)), it follows from claims (i), (iii) above that $u^{(j)} \in C^\infty(\mathbb{R}^3 \times [t_j, t_{j+1}]; \mathbb{R}^3)$ is divergence free and satisfies the Navier–Stokes inequality

$$\partial_t \left| u^{(j)} \right|^2 \leq -u^{(j)} \cdot \nabla \left(\left| u^{(j)} \right|^2 + 2p^{(j)} \right) + 2\nu\, u^{(j)} \cdot \Delta u^{(j)} \qquad (4.21)$$

in $\mathbb{R}^3 \times [t_j, t_{j+1}]$ for all $\nu \in [0, \nu_0]$, where $p^{(j)}$ is the pressure function corresponding to $u^{(j)}$ (recall that $C^\infty(\mathbb{R}^3 \times [a, b]; \mathbb{R}^3)$ denotes the space of vector functions that are infinitely differentiable on $\mathbb{R}^3 \times (a - \eta, b + \eta)$ for some $\eta > 0$). Moreover, although the rescaling used in (4.19) might seem different from the one in (4.20), note that by (4.20), the rescaling used in (4.18) and (i), (ii) we obtain

$$\operatorname{supp} u^{(j)}(t) = \bigcup_{m \in M(j)} \Gamma_m(G), \qquad t \in [t_j, t_{j+1}] \qquad (4.22)$$

(cf. the previous relation (3.9)) and

$$\left| u^{(j)}(x, t_j) \right| = \tau^{-j} \sum_{m \in M(j)} h_0(R^{-1}(\Gamma_m^{-1}(x))),$$
$$\left| u^{(j)}(x, t_{j+1}) \right|^2 > \tau^{-2j} \sum_{m \in M(j)} h_T(R^{-1}(\Gamma_m^{-1}(x)))^2 - \tau^{-2j}\theta. \quad x \in \mathbb{R}^3. \qquad (4.23)$$

The last two lines can be used to show that

$$\left| u^{(j)}(x, t_j) \right| \leq \left| u^{(j-1)}(x, t_j) \right|, \qquad x \in \mathbb{R}^3, j \geq 1, \qquad (4.24)$$

in a similar way as (3.10). Indeed, in order to see this note that this inequality is nontrivial only for $x \in \bigcup_{m \in M(j)} \Gamma_m(G)$, and so let $j \geq 1$, $m \in M(j)$ be such that

$x = \Gamma_m(y)$ for some $y \in G$. Then, in the light of (4.7), we see that $\Gamma_{\tilde{m}}^{-1}(x) \notin G$ for any $\tilde{m} \in M(j)$, $\tilde{m} \neq m$, and so the first line of (4.23) becomes simply

$$\left|u^{(j)}(x, t_j)\right| = \tau^{-j} h_0(R^{-1}(y)). \tag{4.25}$$

Furthermore, letting $\overline{m} \in M(j-1)$ be the sub-multiindex of m, that is $m = (\overline{m}, m_j)$ for some $m_j \in \{1, \ldots, M\}$, we see that

$$x = \Gamma_{\overline{m}}\left(\Gamma_{m_j}(y)\right).$$

This means that, at $(j-1)$-th step (that is for $t \in [t_{j-1}, t_j]$) x was an element of $\Gamma_{\overline{m}}(G)$ (and at time t_j this component of supp $u^{(j-1)}$ will divide into M disjoint copies, $\{\Gamma_{\overline{m},n}(G)\}_{n=1,\ldots,M}$, which will become M out of M^j components of supp $u^{(j)}$ (see (4.22)); and among these copies x belongs to $\Gamma_m(G)$). Therefore, as in (4.25) above we see that the second line of (4.23) (taken for $j-1$) is simply

$$\begin{aligned}
\left|u^{(j-1)}(x, t_j)\right|^2 &> \tau^{-2(j-1)} h_T(R^{-1}(\Gamma_{\overline{m}}^{-1}(x)))^2 - \tau^{-2(j-1)}\theta \\
&= \tau^{-2(j-1)} h_T(R^{-1}(\Gamma_{m_j}(y)))^2 - \tau^{-2(j-1)}\theta. \\
&= \tau^{-2(j-1)} h_{2,T}(R^{-1}(\Gamma_{m_j}(y)))^2 - \tau^{-2(j-1)}\theta,
\end{aligned}$$

where, in the last equality, we used the fact that $h_{1,T}(R^{-1}(\Gamma_{m_j}(y))) = 0$ (recall (4.3) gives $R^{-1}(\Gamma_{m_j}(y)) \in \overline{U_2}$). From this and (4.25), we obtain (4.24) by an easy calculation,

$$\begin{aligned}
\left|u^{(j-1)}(x, t_j)\right|^2 &> \tau^{-2(j-1)} h_{2,T}(R^{-1}(\Gamma_{m_j}(y)))^2 - \tau^{-2(j-1)}\theta \\
&> \tau^{-2j}\left(f_1(R^{-1}(y)) + f_2(R^{-1}(y))\right)^2 \\
&= \tau^{-2j} h_0(R^{-1}(y))^2 \\
&= \left|u^{(j)}(x, t_j)\right|^2,
\end{aligned}$$

where we used (4.17) in the second inequality.

Hence, letting

$$u(t) := \begin{cases} u^{(j)}(t) & \text{if } t \in [t_j, t_{j+1}) \text{ for some } j \geq 0, \\ 0 & \text{if } t \geq T_0, \end{cases} \tag{4.26}$$

where $T_0 := \lim_{j \to \infty} t_j = T/(1 - \tau^2)$ (as previously), we obtain a solution of Theorem 4.1. Indeed, that u is a weak solution of the NSI follows as in the case of Theorem 3.1 (note that, in order to obtain the required regularity $\sup_{t>0}\|u\| < \infty$, $\nabla u \in L^2(\mathbb{R}^3 \times (0, \infty))$ it suffices to replace "τ" by "$M\tau$" and $\sup_{t \in [t_0, t_1]}\|u^{(0)}(t)\|$, $\int_{t_0}^{t_1}\|\nabla u^{(0)}(t)\|^2 dt$ by $\mathcal{C}$ (from Proposition 4.3 (iv); recall also the shorthand notation $\|\cdot\| \equiv \|\cdot\|_{L^2(\mathbb{R}^3)}$)) in the calculations (3.11), (3.12).

Furthermore, the singular set of $\mathfrak{u}$ is

$$S \times \{T_0\} = \left(\bigcap_{j \geq 0} \bigcup_{m \in M(j)} \Gamma_m(G) \right) \times \{T_0\}.$$

Indeed, (4.22) shows that the support of $\mathfrak{u}(t)$ consists of M^j components for $t \in [t_j, t_{j+1})$ and that it shrinks to the Cantor set S as $t \to T_0^-$. That $\mathfrak{u}$ is unbounded in any neighbourhood V of any point $(y, T_0) \in S \times \{T_0\}$ follows from Proposition 4.3 (ii) and (4.20), which show that the magnitude of $\mathfrak{u}$ grows uniformly on each component of its support; in other words given any positive number $\mathcal{N}$ let $y \in G$ be any point such that $h_0(R^{-1}(y)) > 0$ and let $j \geq 0$, $m \in M(j)$ be such that

$$\Gamma_m(G) \times [t_j, T_0] \subset V \quad \text{and} \quad \tau^{-j} \geq \mathcal{N}/h_0(R^{-1}(y)).$$

Then, by (4.25),

$$|\mathfrak{u}(\Gamma_m(y), t_j)| = \tau^{-j} h_0(R^{-1}(y)) \geq \mathcal{N}.$$

Thus $S \times \{T_0\}$ is a singular set of $\mathfrak{u}$ whose Hausdorff dimension is greater than ξ (due to (4.10)).

4.3 Proof of Proposition 4.3

Here, we prove Proposition 4.3, which concludes the proof of Theorem 4.1 (given the geometric arrangement, which we discuss in Sect. 4.5). Fix $j \geq 0$. We will write, for brevity, $v = v^{(j)}, \overline{p} = \overline{p}^{(j)}$.

Step 1. We renumber the functions $h_i(\pi_m^{-1}(\tau^j x_1), x_2, t)$.

For brevity we let $\mathfrak{M} := M^j$, identify each multiindex $m \in M(j)$ with an integer $m \in \{1, \ldots, \mathfrak{M}\}$, and let

$$h_i^{\mathrm{m}}(x_1, x_2, t) := h_i(\pi_m^{-1}(\tau^j x_1), x_2, t), \qquad i = 1, 2 \tag{4.27}$$

(recall that $h_t = h_1(\cdot, t) + h_2(\cdot, t)$), and

$$\begin{cases} f_i^{\mathrm{m}}(x_1, x_2) := f_i(\pi_m^{-1}(\tau^j x_1), x_2), \\ v_i^{\mathrm{m}}(x_1, x_2) := v_i(\pi_m^{-1}(\tau^j x_1), x_2), \\ \phi_i^{\mathrm{m}}(x_1, x_2) := \phi_i(\pi_m^{-1}(\tau^j x_1), x_2), \\ U_i^{\mathrm{m}} := \{(x_1, x_2) \colon (\pi_m^{-1}(\tau^j x_1), x_2) \in U_i\} \end{cases} \qquad i = 1, 2.$$

Then $h_1^m, h_2^m \in C^\infty(P \times (-\delta, T + \delta); [0, \infty))$,

(v_i^m, f_i^m, ϕ_i^m) and $(v_i^m, h_{i,t}^m, \phi_i^m)$ are structures on U_i^m for $t \in (-\delta, T + \delta), i = 1, 2,$

and

$$h_t^{(j)} = \sum_{m=1}^{\mathfrak{M}} \left(h_{1,t}^m + h_{2,t}^m \right).$$

Moreover,

$$\text{supp}\,(h_{1,t}^m + h_{2,t}^m) = \overline{U_1^m} \cup \overline{U_2^m} =: K^m$$

and the sets K^m are pairwise disjoint translates of $\overline{U_1} \cup \overline{U_2}$ in the x_1 direction, such that the distance between any K^m and K^n for $m, n \in \{1, \ldots, \mathfrak{M}\}$, $n \neq m$, is at least $\tau^{-1}\zeta$ (just as each element of the union $\bigcup_{m \in M(j)} \Gamma_m(G)$ is separated from the rest by at least $\tau^{j-1}\zeta$, see the comments preceding (4.10)). Furthermore, we can assume that the bijection $m \longleftrightarrow \mathfrak{m}$ is such that K^{m+1} is a positive translate of K^m in the x_1 direction, that is

$$K^{m+1} = K^m + (a_m, 0) \quad \text{for some } a_m > 0, \quad m = 1, \ldots, \mathfrak{M} - 1. \qquad (4.28)$$

For such a bijection

$$\text{dist}(K^n, K^m) \geq |n - m|\tau^{-1}\zeta, \qquad n, m = 1, \ldots, \mathfrak{M}. \qquad (4.29)$$

Step 2. We introduce the modification $q_{i,t}^{m,k}$ of $h_{i,t}^m$.
    Let

$$\left(q_{i,t}^{m,k} \right)^2 := \left(h_{i,0}^m \right)^2 - 2t\delta\phi_i^m - \int_0^t a_i^{m,k}(s)v_i^m \cdot \left(\nabla (h_{i,s}^m)^2 + 2 \sum_{l=1,2} \sum_{n=1}^{\mathfrak{M}} \nabla p[a_l^{n,k}(s)v_l^n, h_{l,s}^n] \right) ds, \qquad (4.30)$$

$i = 1, 2, \quad k \in \mathbb{N}, \quad m = 1, \ldots, \mathfrak{M}, \quad$ where $a_i^{m,k} \in C^\infty(\mathbb{R}; [-1, 1]), \quad i = 1, 2, \quad m = 1, \ldots, M$ are the oscillatory functions which we construct below. Observe that this is a natural extension of the idea from Sect. 3.3.1 to the case of $\mathfrak{M}$ pairs U_1^m, U_2^m (rather than a single pair U_1, U_2, which was the case previously). Note that such a definition gives

$$\partial_t \left(q_{i,t}^{m,k} \right)^2 = -2\delta\phi_i^m + a_i^{m,k}(t)v_i^m \cdot \left(\nabla (h_{i,t}^m)^2 + 2 \sum_{l=1,2} \sum_{n=1}^{\mathfrak{M}} \nabla p[a_l^{n,k}(t)v_l^n, h_{l,t}^n] \right). \qquad (4.31)$$

As in (3.78), we will construct the oscillatory processes $a_1^{m,k}, a_2^{m,k} \in C^\infty(\mathbb{R}; [-1, 1])$ in such a way that for each multiindex $\alpha = (\alpha_1, \alpha_2)$

$$
\begin{cases}
q_{i,t}^{m,k} \to h_{i,t}^m \\
\text{and} \\
D^\alpha q_{i,t}^{m,k} \to D^\alpha h_{i,t}^m
\end{cases}
\qquad \text{uniformly in } P \times [0, T], i = 1, 2, \mathfrak{m} \in \{1, \dots, \mathfrak{M}\}.
$$

(4.32)

As in Sect. 3.3.1, we obtain that for $i = 1, 2$ and sufficiently large k

$$(a_i^{m,k}(t)v_i^m, q_{i,t}^{m,k}, \phi_i^m) \text{ is a structure on } U_i^m \text{ for } t \in (-\delta_k, T + \delta_k),$$

and that

$$q_i^{m,k} \in C^\infty(P \times (-\delta_k, T + \delta_k); [0, \infty)).$$

Finally let

$$v(x, t) := \sum_{m=1}^{\mathfrak{M}} \left(u[a_1^{m,k}(t)v_1^m, q_{1,t}^{m,k}](x) + u[a_2^{m,k}(t)v_2^m, q_{2,t}^{m,k}](x) \right). \qquad (4.33)$$

Step 3. We verify that v satisfies the claims of the proposition.

We will now show that (given the existence of the oscillatory processes $a_1^{m,k}, a_2^{m,k}$, which we construct in Sect. 4.4) the function (4.33) is a solution of Proposition 4.3.

Claim (i) is trivial and so is claim (ii) given k large enough such that

$$\left| (q_{i,t}^{m,k})^2 - (h_{i,t}^m)^2 \right| \leq \theta/2 \qquad \text{in } P, t \in [0, T], i = 1, 2.$$

As for claim (iii) (the Navier–Stokes inequality) note that (as previously) since $v^{(j)}$ is axisymmetric, it is equivalent to

$$\partial_t |v(x, 0, t)|^2 \leq -v(x, 0, t) \cdot \nabla \left(|v(x, 0, t)|^2 + 2\overline{p}(x, 0, t) \right) + 2\nu \, v(x, 0, t) \cdot \Delta v(x, 0, t),$$

where $\nu \in [0, \nu_0]$, $x \in P$, $t \in [0, T]$ and $\overline{p}$ is the pressure function corresponding to v, that is

$$\overline{p}(t) = \sum_{m=1}^{\mathfrak{M}} \left(p^*[a_1^{m,k}(t)v_1^m, q_{1,t}^{m,k}] + p^*[a_2^{m,k}(t)v_2^m, q_{2,t}^{m,k}] \right) \qquad (4.34)$$

(recall (3.33) and Lemma 3.3 (iii)), which in particular (i.e. restricting ourselves to P) means that

$$\overline{p}(x,0,t) = \sum_{m=1}^{\mathfrak{M}} \left(p[a_1^{m,k}(t)v_1^m, q_{1,t}^{m,k}](x) + p[a_2^{m,k}(t)v_2^m, q_{2,t}^{m,k}](x) \right), \quad x \in P.$$

As in Sect. 3.3.2 we fix $x \in P, t \in [0, T]$ and consider two cases.

Case 1. $\phi_1^m(x) + \phi_2^m(x) < 1$ for all $m \in \{1, \ldots, \mathfrak{M}\}$.

For such x we have $v_1^m(x) = v_2^m(x) = 0$ (from the elementary properties of structures, recall Definition 3.4) and the Navier–Stokes inequality follows trivially for all k by writing

$$\partial_t |v(x,0,t)|^2 = \sum_{m=1}^{\mathfrak{M}} \left(\partial_t q_{1,t}^{m,k}(x)^2 + \partial_t q_{2,t}^{m,k}(x)^2 \right)$$

$$= -2\delta \sum_{m=1}^{\mathfrak{M}} (\phi_1^m(x) + \phi_2^m(x))$$

$$\leq 0$$

$$\leq -v(x,0,t) \cdot \nabla \left(|v(x,0,t)|^2 + 2\overline{p}(x,0,t) \right) + 2\nu\, v(x,0,t) \cdot \Delta v(x,0,t),$$

where we used (3.43) and (3.44) in the last step.

Case 2. $\phi_1^m(x) + \phi_2^m(x) = 1$ for some $m \in \{1, \ldots, \mathfrak{M}\}$.

In this case, we need to use the convergence (4.32) with k sufficiently large such that

$$|v_i^m| \left(\left| \nabla(q_{i,t}^{m,k})^2 - \nabla(h_{i,t}^m)^2 \right| + 2 \sum_{n=1}^{\mathfrak{M}} \sum_{l=1,2} \left| \nabla p[a_l^{n,k}(t)v_l^n, q_{l,t}^{n,k}] - \nabla p[a_l^{n,k}(t)v_l^n, h_{l,t}^n] \right| \right) \leq \delta/2$$

$$(4.35)$$

in P (see Lemma 3.17 for a verification that (4.32) is sufficient for the convergence of the pressure functions) and

$$\nu_0 \left| u[a_i^{m,k}(t)v_i^m, q_{i,t}^{m,k}] \cdot \Delta u[a_i^{m,k}(t)v_i^m, q_{i,t}^{m,k}] \right|$$

$$\leq \nu_0 \left| u[a_i^{m,k}(t)v_i^m, h_{i,t}^m] \cdot \Delta u[a_i^{m,k}(t)v_i^m, h_{i,t}^m] \right| + \delta/8 \leq \delta/4$$

$$(4.36)$$

in $\mathbb{R}^3$ (see Lemma 3.18 for a verification that (4.32) is sufficient for the the first inequality), for $t \in [0, T], i = 1, 2$, where we used (3.71) in the last inequality. Recall that $\delta > 0$ has been fixed below (4.16). We obtain

$$\partial_t |v(x,0,t)|^2 = \partial_t q_{1,t}^{m,k}(x)^2 + \partial_t q_{2,t}^{m,k}(x)^2$$

$$= -2\delta - \left(a_1^{m,k}(t)v_1^m(x) + a_2^{m,k}(t)v_2^m(x)\right) \cdot \nabla \left(\left(h_{1,t}^m(x)\right)^2 + \left(h_{2,t}^m(x)\right)^2 \sum_{n=1}^{\mathfrak{M}}\right.$$

$$\left. + 2\sum_{n=1}^{\mathfrak{M}}\left(p[a_1^{n,k}(t)v_1^n, h_{1,t}^n](x) + p[a_2^{n,k}(t)v_2^n, h_{2,t}^n](x)\right)\right)$$

$$\leq -\delta - \left(a_1^{m,k}(t)v_1^m(x) + a_2^{m,k}(t)v_2^m(x)\right) \cdot \nabla \left(q_{1,t}^{m,k}(x)^2 + q_{2,t}^{m,k}(x)^2 \sum_{n=1}^{\mathfrak{M}}\right.$$

$$\left. + 2\sum_{n=1}^{\mathfrak{M}}\left(p[a_1^{n,k}(t)v_1^n, q_{1,t}^{n,k}](x) + p[a_2^{n,k}(t)v_2^n, q_{2,t}^{n,k}](x)\right)\right)$$

$$= -\delta - v_1(x,0,t)\partial_{x_1}\left(|v(x,0,t)|^2 + 2\overline{p}(x,0,t)\right)$$

$$- v_2(x,0,t)\partial_{x_2}\left(|v(x,0,t)|^2 + 2\overline{p}(x,0,t)\right),$$

and so, recalling that $\partial_{x_3}|v(x,0,t)|^2 = \partial_{x_3}\overline{p}(x,0,t) = 0$ (as a property of axisymmetric functions, see (3.31) and (3.36)),

$$\partial_t |v(x,0,t)|^2 \leq -\delta - v(x,0,t) \cdot \nabla \left(|v(x,0,t)|^2 + 2\overline{p}(x,0,t)\right)$$

$$\leq 2\nu\, v(x,0,t) \cdot \Delta v(x,0,t) - v(x,0,t) \cdot \nabla \left(|v(x,0,t)|^2 + 2\overline{p}(x,0,t)\right)$$

for all $\nu \in [0, \nu_0]$, where we used (4.36) in the last step.

It remains to verify (iv). For this note that

$$|v(x,t)| = \sum_{m=1}^{\mathfrak{M}} \left|q_{1,t}^{m,k}(R^{-1}x) + q_{2,t}^{m,k}(R^{-1}x)\right|$$

(recall that $\{q_{i,t}^{m,k}\}_{i=1,2,\, m=1,\ldots,\mathfrak{M}}$ have disjoint supports U_i^m, respectively), and thus, in the view of (4.32), for sufficiently large k

$$|v(x,t)| \leq \sum_{m=1}^{\mathfrak{M}} \left|h_{1,t}^m(R^{-1}x) + h_{2,t}^m(R^{-1}x)\right| + 1$$

$$\leq \sup_{s\in[0,T]} \|h_{1,s} + h_{2,s}\|_{L^\infty} + 1 \tag{4.37}$$

$$= \mathcal{C},$$

since the functions $h_{1,t}^m + h_{2,t}^m$ have disjoint supports K^m ($m = 1, \ldots, \mathfrak{M}$). Here we write $\mathcal{C}$ for a constant that is independent of j whose value might change from line to line (this should not be confused with C, which is a constant related to the decay of the pressure function and was fixed above Sect. 3.4.1). Hence, since

$\operatorname{supp} v(t) = \bigcup_{m=1}^{\mathfrak{M}} R(K^m)$ consists of $\mathfrak{M}$ copies of $R(\overline{U_1} \cup \overline{U_2})$ (recall (4.28)), we obtain, by Hölder's inequality, that

$$\|v(t)\|_{L^2} \leq \mathfrak{M}\mathcal{C}, \qquad \text{for } t \in [0, T]$$

for some $\mathcal{C} > 0$.

As in (4.37) we have for sufficiently large k

$$\|q_{1,t}^{m,k} + q_{2,t}^{m,k}\|_{W^{1,\infty}} \leq \|h_{1,t}^m + h_{2,t}^m\|_{W^{1,\infty}} + 1,$$

and so, applying (3.46), we obtain

$$
\begin{aligned}
|\nabla v(x,t)| &\leq \sum_{m=1}^{\mathfrak{M}} \left| \nabla u[a_1^{m,k}(t)v_1^m, q_{1,t}^{m,k}](x) + \nabla u[a_2^{m,k}(t)v_2^m, q_{2,t}^{m,k}](x) \right| \\
&\leq \max_{m \in \{1,\ldots,\mathfrak{M}\}} \mathcal{C}(\|v_1^m + v_2^m\|_{W^{1,\infty}}, \|q_{1,t}^{m,k} + q_{2,t}^{m,k}\|_{W^{1,\infty}}) \\
&\leq \max_{m \in \{1,\ldots,\mathfrak{M}\}, s \in [0,T]} \mathcal{C}(\|v_1^m + v_2^m\|_{W^{1,\infty}}, \|h_{1,s}^m + h_{2,s}^m\|_{W^{1,\infty}} + 1) \\
&= \max_{s \in [0,T]} \mathcal{C}(\|v_1 + v_2\|_{W^{1,\infty}}, \|h_{1,s} + h_{2,s}\|_{W^{1,\infty}} + 1) \\
&= \mathcal{C},
\end{aligned}
\tag{4.38}
$$

and therefore

$$\int_0^T \|\nabla v(t)\|_{L^2}^2 \, dt \leq \mathfrak{M}\mathcal{C},$$

as required.

4.4 The New Oscillatory Processes

Here, we prove the existence of oscillatory processes $a_1^{m,k}, a_2^{m,k} \in C^\infty(\mathbb{R}; [-1,1])$ which give the convergence (4.32). The construction of such oscillatory processes is a natural extension of the construction of the processes a_1^k, a_2^k from Sect. 3.3.3 to the case of $\mathfrak{M}$ pairs U_1^m, U_2^m (and the corresponding structures, $m = 1, \ldots, \mathfrak{M}$). In particular, we will use the following sharper version of Theorem 3.9.

Theorem 4.4 *For each $k \geq 1$, $m = 1, \ldots, \mathfrak{M}$ there exists a pair of functions $a_1^{m,k}, a_2^{m,k} \in C^\infty(\mathbb{R}; [-1,1])$, $i = 1, 2$, such that*

$$
\int_0^t a_i^{m,k}(s) \left(G_i^m(x,s) + \sum_{l=1,2} \sum_{n=1}^{\mathfrak{M}} F_{i,l}^{m,n}\left(x, s, a_j^{n,k}(s)\right) \right) ds
$$

$$
\xrightarrow{k \to \infty}
\begin{cases}
\frac{1}{2} \int_0^t \left(F_{2,1}^{m,m}(x,s,1) - F_{2,1}^{m,m}(x,s,0) \right) ds & i = 2, \\
0 & i = 1
\end{cases}
\tag{4.39}
$$

uniformly in $(x, t) \in P \times [0, T]$, $\mathfrak{m} = 1, \ldots, \mathfrak{M}$ *for any bounded and uniformly continuous functions*

$$G_i^{\mathfrak{m}} : P \times [0, T] \to \mathbb{R}, \qquad F_{i,l}^{\mathfrak{m},\mathfrak{n}} : P \times [0, T] \times [-1, 1] \to \mathbb{R},$$

$i, l = 1, 2$, $\mathfrak{m}, \mathfrak{n} = 1, \ldots, \mathfrak{M}$ *satisfying*

$$F_{i,l}^{\mathfrak{m},\mathfrak{n}}(x, t, -1) = F_{i,l}^{\mathfrak{m},\mathfrak{n}}(x, t, 1) \quad \text{for } x \in P, t \in [0, T], i, l = 1, 2, \mathfrak{m}, \mathfrak{n} = 1, \ldots, \mathfrak{M}.$$

Note that, as in Sect. 3.3.3, this theorem gives (4.32) simply by taking

$$G_i^{\mathfrak{m}}(x, t) := v_i^{\mathfrak{m}}(x) \cdot \nabla(h_i^{\mathfrak{m}}(x, t))^2,$$
$$F_{i,l}^{\mathfrak{m},\mathfrak{n}}(x, t, a) := 2v_i^{\mathfrak{m}}(x) \cdot \nabla p[av_l^{\mathfrak{n}}, h_{l,t}^{\mathfrak{n}}](x)$$

(recall $F_{i,l}^{\mathfrak{m},\mathfrak{n}}(x, t, -1) = F_{i,l}^{\mathfrak{m},\mathfrak{n}}(x, t, 1)$ by the property $p[v, f] = p[-v, f]$, see Lemma 3.3 (i)) and by taking

$$G_i^{\mathfrak{m}}(x, t) := D^{\alpha} \left(v_i^{\mathfrak{m}}(x) \cdot \nabla(h_i^{\mathfrak{m}}(x, t))^2 \right),$$
$$F_{i,l}^{\mathfrak{m},\mathfrak{n}}(x, t, a) := D^{\alpha} \left(2v_i^{\mathfrak{m}}(x) \cdot \nabla p[av_l^{\mathfrak{n}}, h_{l,t}^{\mathfrak{n}}](x) \right)$$

for any given multiindex $\alpha = (\alpha_1, \alpha_2)$.

In order to see that the theorem above is a sharpening of Theorem 3.9, recall that the role of the processes a_1^k, a_2^k (given by Theorem 3.9) was (in a sense) to "pick" (among all influences of the set U_i on the set U_j, $i, j \in \{1, 2\}$) only the influence of U_1 on U_2 (recall the comments following Theorem 3.9). Here, instead of a pair U_1, U_2 we have to deal with $\mathfrak{M}$ pairs $U_1^{\mathfrak{m}}$, $U_2^{\mathfrak{m}}$ ($\mathfrak{m} = 1, \ldots, \mathfrak{M}$) and the role of the processes $a_1^{\mathfrak{m},k}$, $a_2^{\mathfrak{m},k}$ is to "pick" (among all influences of $U_i^{\mathfrak{n}}$ on $U_l^{\mathfrak{m}}$, $i, l \in \{1, 2\}$, $\mathfrak{n}, \mathfrak{m} \in \{1, \ldots, \mathfrak{M}\}$) only the influence of $U_1^{\mathfrak{m}}$ on $U_2^{\mathfrak{m}}$ for all $\mathfrak{m} \in \{1, \ldots, \mathfrak{M}\}$ (that is for each pair pick only the influence of the first set on the second one). Thus, recalling that the choice of the processes a_1^k, a_2^k (in Sect. 3.3.3) was based on the "basic processes" b_1, b_2 (recall (3.86)) having the simple integral property (3.87), we can obtain the processes $a_1^{\mathfrak{m},k}$, $a_2^{\mathfrak{m},k}$ by finding processes $b_1^{(\mathfrak{m})}$, $b_2^{(\mathfrak{m})}$, $\mathfrak{m} = 1, \ldots, \mathfrak{M}$ such that an analogous property holds:

$$\int_0^T b_i^{(\mathfrak{n})}(s) f\left(b_l^{(\mathfrak{m})}(s) \right) ds = \begin{cases} \frac{T}{2}(f(1) - f(0)) & (i, l) = (2, 1), \mathfrak{m} = \mathfrak{n}, \\ 0 & \text{otherwise} \end{cases} \tag{4.40}$$

for any $f : [-1, 1] \to \mathbb{R}$ such that $f(-1) = f(1)$. Such processes can be obtained by letting $b_1^{(1)} := b_1, b_2^{(1)} := b_2$ and letting $b_i^{(\mathfrak{m})}$ have 4 times higher frequency than $b_i^{(\mathfrak{m}-1)}$, $i = 1, 2$, $\mathfrak{m} \in \{2, \ldots, \mathfrak{M}\}$, that is letting

$$b_1^{(\mathfrak{m})}(t) := b_1(4^{\mathfrak{m}-1}t), \qquad b_2^{(\mathfrak{m})}(t) := b_2(4^{\mathfrak{m}-1}t) \tag{4.41}$$

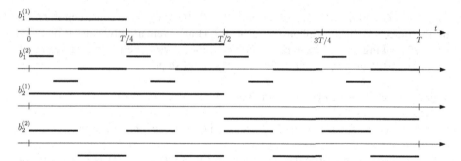

Fig. 4.3 The processes $b_1^{(m)}, b_2^{(m)}$, m $= 1, \ldots, \mathfrak{M}$. Here $\mathfrak{M} = 2$

where we have also extended b_1, b_2 T-periodically to the whole line, see Fig. 4.3. Analogously as in Sect. 3.3.3 the convergence (4.39) can be obtained by letting, for each k, $b_1^{m,k}, b_2^{m,k}$ (m $= 1, \ldots, \mathfrak{M}$) be oscillations of the above form with frequency increasing with k, that is

$$b_1^{m,k}(t) := b_1^{(m)}(kt) = b_1(k4^{m-1}t), \quad b_2^{m,k}(t) := b_2^{(m)}(kt) = b_2(k4^{m-1}t). \quad (4.42)$$

We omit the detailed calculation.

Finally, as in Sect. 3.3.3, the smoothness of the processes can be obtained by smooth approximation of the processes $b_1^{m,k}, b_2^{m,k}$, that is by letting $a_1^{m,k}, a_2^{m,k} \in C^\infty(\mathbb{R}; [-1, 1])$ be such that

$$\left| \left\{ t \in [0, T] : a_i^{m,k}(t) \neq b_i^{m,k}(t) \right\} \right| \leq \frac{1}{k}, \quad i = 1, 2, \text{m} = 1, \ldots, \mathfrak{M}.$$

4.5 The New Geometric Arrangement

In this section, we construct the geometric arrangement as described in Sect. 4.2. That is we need to find $U_1, U_2 \Subset P$ (with disjoint closures) together with the corresponding structures (v_1, f_1, ϕ_1), (v_2, f_2, ϕ_2) and numbers $T > 0$, $\tau \in (0, 1)$, $z = (z_1, z_2, 0) \in \mathbb{R}^3$, $X > 0$, $M \in \mathbb{N}$ such that except for (3.51), (3.52) (which was all that we required in the proof of Theorem 3.1, recall Sect. 3.3) we also have (4.2), (4.3) and (4.12), that is $\{\Gamma_n(G)\}_{n=1,\ldots,M}$ is a family of pairwise disjoint subsets of G_2 (recall $G = G_1 \cup G_2 = R(\overline{U_1}) \cup R(\overline{U_2})$),

$$\tau^\xi M \geq 1, \quad \tau M < 1$$

and

$$f_2^2(y_n) + T v_2(y_n) \cdot F[v_1, f_1](y_n) > \tau^{-2} \left(f_1(R^{-1}x) + f_2(R^{-1}x) \right)^2$$

for all $x \in G$ and $n = 1, \ldots, M$, where $y_n = R^{-1}(\Gamma_n(x))$. The construction builds on the objects defined previously (in Sect. 3.4) and, remarkably, can be obtained simply by taking $\varepsilon > 0$ (the main parameter of the previous geometric arrangement, recall (3.110)) smaller, which we present in several steps.

Step 1. We recall some objects from Sect. 3.4.

Let
$$U, v, f, \phi, F, A, B, C, D, \kappa \quad \text{and } a', r', s', a'', r'', s'', H, E$$

be as in Sect. 3.4. In particular, U is a rectangle in P, (v, f, ϕ) is a structure on U, $F = F[v, f]$ is the corresponding pressure interaction function, the constants $A, B, C, D \in \mathbb{R}$ are given by the properties of the pressure interaction function F (recall Lemma 3.6), $\kappa = 10^4 C/D \geq 10^4$ (recall (3.49)), the numbers a', r', s', a'', r'', s'' define the copies $U^{a',r'}$, $U^{a'',r''}$ of U (and the copies of the corresponding structures) in a way that the joint pressure interaction function $H = F + F^{a',r',s'} + F^{a'',r'',s''}$ has certain decay and certain behaviour on the x_1 axis (that is (i)–(iii) from Sect. 3.4.2 hold), and $E > 0$ is sufficiently small such that the strip $0 < x_2 < E$ is disjoint with $U \cup U^{a',r'} \cup U^{a'',r''}$ and H enjoys certain properties in this strip (that is (iv)–(vi) from Sect. 3.4.2 hold).

Step 2. We consider disjoint copies of $U \cup U^{a',r'} \cup U^{a'',r''}$ in the x_1 direction.

That is, we let $X > 0$ be sufficiently large so that

$$X > \text{diam}\left(U \cup U^{a',r'} \cup U^{a'',r''}\right), \qquad X > 4|A|,$$

$$2CX^{-4} \sum_{k \in \mathbb{Z}} \left(|k| - \frac{1}{2}\right)^{-4} < 0.01B, \qquad \text{and} \qquad X > 2\kappa E, \tag{4.43}$$

and consider the collection of copies of $U \cup U^{a',r'} \cup U^{a'',r''}$:

$$\left\{ U^{nX,1} \cup U^{a'+nX,r'} \cup U^{a''+nX,r''} \right\}_{n \in \mathbb{Z}}, \tag{4.44}$$

together with the structures that are the corresponding translations by $(nX, 0)$ of

$$(v, f, \phi) + \left(v^{a',r',s'}, f^{a',r',s'}, \phi^{a',r'}\right) + \left(v^{a'',r'',s''}, f^{a'',r'',s''}, \phi^{a'',r''}\right),$$

recall (3.107) (see Fig. 4.4). The role of X is to separate these copies (and the corresponding structures) sufficiently far from each other. In particular, we see that they have disjoint closures by the first inequality in (4.43). Note also that since each of $U^{nX} \cup U^{a'+nX,r'} \cup U^{a''+nX,r''}$, $n \in \mathbb{Z}$, is a translation in the x_1 direction of $U \cup U^{a',r'} \cup U^{a'',r''}$, it is disjoint with the strip $\{0 < x_2 < E\}$ (recall (iv) in Sect. 3.4.2), see Fig. 4.4.

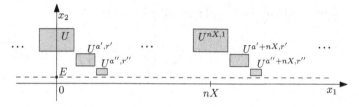

Fig. 4.4 The sets $U^{nX,1} \cup U^{a'+nX,r'} \cup U^{a''+nX,r''}, n \in \mathbb{Z}$

Moreover, note that for each $n \in \mathbb{Z}$

$$H(x_1 - nX, x_2) = \left(F^{nX,1} + F^{a'+nX,r',s'} + F^{a''+nX,r'',s''} \right)(x_1, x_2), \qquad (x_1, x_2) \in \mathbb{R}^2,$$

that is $H(x_1 - nX, x_2)$ is the pressure interaction function corresponding to $U^{nX,1} \cup U^{a'+nX,r'} \cup U^{a''+nX,r''}$ (with the structure as pointed out above). We now show that the choice of X above gives that for each $k \in \mathbb{Z}$ the total pressure interaction of the sets (4.44) with $n \neq k$ (and their structures) is very small near $U^{kX} \cup U^{a'+kX,r'} \cup U^{a''+kX,r''}$, which we make precise in the following lemma.

Lemma 4.5 *Given $x_1 \in \mathbb{R}$ let $k \in \mathbb{Z}$ be such that*

$$|x_1 - kX| = \min_{n \in \mathbb{Z}} |x_1 - nX|.$$

Then
$$\sum_{n \neq k} |H(x_1 - nX, x_2)| < 0.01B, \qquad \text{for all } x_2 \in [0, E).$$

Proof Note that the definition of k means simply that

$$x_1 \in [kX - X/2, kX + X/2].$$

Thus if $n \neq k$ then

$$|x_1 - nX| \geq \left(|n - k| - \frac{1}{2} \right) X,$$

cf. Fig. 4.4. Thus in particular

$$|x_1 - nX| \geq X/2 \geq 2|A|,$$

where we used the fact that $X \geq 4|A|$ (see (4.43)), and so we can use the decay of H (see property (iii) of H) to write

$$|H(x_1 - nX, x_2)| \leq 2C|x_1 - nX|^{-4} \leq 2C \left(|n - k| - \frac{1}{2} \right)^{-4} X^{-4}.$$

Summing up in n and using the third inequality in (4.43), we obtain

$$\sum_{n \neq k} |H(x_1 - nX, x_2)| \leq 2CX^{-4} \sum_{n \neq k} \left(|n - k| - \frac{1}{2}\right)^{-4} \leq 0.01 B. \qquad \square$$

Thus, for any $M \in \mathbb{N}$ the function

$$H^*(x_1, x_2) := \sum_{n=0}^{M-1} H(x_1 - nX, x_2)$$

is the pressure interaction function corresponding to

$$\bigcup_{n=0}^{M-1} \left(U^{nX} \cup U^{a'+nX, r'} \cup U^{a''+nX, r''} \right),$$

and the above lemma and properties (v) and (vi) of H give

(i) $H_1^*(x) \geq -1.02 B$ in the strip $\{0 < x_2 < E\}$,
(ii) $H_1^*(x) \geq 6.98 B$ for $x \in P$ with $|x_1 - A - (m-1)X| < \kappa E$, $0 < x_2 < E$ for any $m = 1, \ldots, M$.

Step 3. We take $\varepsilon > 0$ small, and define v_2, U_2.
    Given $\varepsilon > 0$ let $\tau := 0.48\varepsilon$ and

$$r := E/\varepsilon, \qquad d := \kappa r, \qquad M := 1 + \frac{d}{4X}. \qquad (4.45)$$

(Recall $\kappa \geq 10^4$ (see Step 1).) Note each of r, d, M (is defined in the same way as previously and) is of order ε^{-1}. Let ε be small such that in addition to (3.110) we also have that M is a positive integer and

$$\tau^\xi M \geq 1, \qquad \varepsilon^2 M < \frac{10^{-6} B E^4}{2C}. \qquad (4.46)$$

Note that this gives (4.2), which is clear from the first of the two inequalities above and by writing

$$\tau M = \tau + \frac{\tau d}{4X} = \tau + \frac{0.48\kappa E}{4X} < \tau + \frac{1}{2} < 1,$$

where we used the facts $X > \kappa E/4$ (recall (4.43)) and $\tau < 1/2$ (recall that in fact $\tau < 1/20$ by the first inequality in (3.110)).

Having fixed ε, we let (as previously) v_2 be given by Lemma 3.11 and the sets $U_2, BOX, RECT, SBOX$ be defined as in (3.113), that is,

$$U_2 := (-d, d) \times (0.005\varepsilon r, r) \setminus [-(d - r), d - r] \times [\varepsilon r, r/10],$$
$$BOX := [-d, d] \times [0, r],$$
$$SBOX := [A - \kappa E, A + \kappa E] \times [0.02\varepsilon r, 0.98\varepsilon r],$$
$$RECT := [-(d - r), d - r] \times [0.02\varepsilon r, 0.98\varepsilon r].$$

Note that U_2 encompasses the union of all M copies of $U \cup U^{a',r'} \cup U^{a'',r''}$, that is

$$\bigcup_{n=0}^{M-1} \left(U^{nX,1} \cup U^{a'+nX,r'} \cup U^{a''+nX,r''} \right) \subset (-(d - r), d - r) \times (\varepsilon r, r/10) \quad (4.47)$$

(see Fig. 4.6), which can be verified in the same way as (3.115), except for the use of the inequality $d - r > \text{diam}\left(U \cup U^{a',r'} \cup U^{a'',r''} \right)$, which can be improved by using the fifth inequality in (3.110),

$$d > 2\text{diam}\left(U \cup U^{a',r'} \cup U^{a'',r''} \right),$$

to give

$$d - r = \left(\frac{d}{2} - r \right) + \frac{d}{2}$$
$$> \frac{d}{4} + \text{diam}\left(U \cup U^{a',r'} \cup U^{a'',r''} \right)$$
$$= (M - 1)X + \text{diam}\left(U \cup U^{a',r'} \cup U^{a'',r''} \right),$$

where we also used the fact that $d > 4r$ (recall (4.45) above). Thus we obtain (4.47).
    Let

$$SBOX_m := SBOX + (m - 1)(X, 0), \quad m = 1, \ldots, M, \quad (4.48)$$

and observe that $\{SBOX_m\}_{m=1}^M$ is a family of pairwise disjoint subsets of $RECT$ (cf. Fig. 4.5). Indeed, the disjointness follows from the fact that $X > 2\kappa E$ (recall the last inequality in (4.43)), the inclusion $SBOX_1 \subset RECT$ follows as previously (recall the comment following (3.113)) and the inclusion $SBOX_M \subset RECT$ follows by writing

$$(M - 1)X + A + \kappa E = \frac{d}{4} + A + \kappa E < \frac{d}{4} + (d - r)/2 < d - r,$$

where we used the second inequality in (3.110) and the fact that $d > 2r$ (recall (3.109)).

Let

$$a := -\kappa r/\varepsilon, \quad \frac{s^2}{r} := 1.04 \left(-\frac{a}{r}\right)^4 B/D$$

(as previously, see (3.117) and recall that then Lemma 3.12 gives

$$1.03 B \le F_1^{a,r,s} \le 1.05 B \quad \text{and} \quad |F_2^{a,r,s}| \le 0.01 \varepsilon B \quad \text{in } BOX. \tag{4.49}$$

Step 4. We define U_1, its structure (v_1, f_1, ϕ_1), and show the lower bound $v_2 \cdot F[v_1, f_1] \ge -1.1 \varepsilon B$.

Letting

$$U_1 := \bigcup_{n=0}^{M-1} \left(U^{nX,1} \cup U^{a'+nX,r'} \cup U^{a''+nX,r''} \right) \cup U^{a,r},$$

and

$$f_1 := \sum_{n=0}^{M-1} \left(f^{nX,1,1} + f^{a'+nX,r',s'} + f^{a''+nX,r'',s''} \right) + f^{a,r,s},$$

$$v_1 := \sum_{n=0}^{M-1} \left(v^{nX,1,1} + v^{a'+nX,r',s'} + v^{a''+nX,r'',s''} \right) + v^{a,r,s},$$

$$\phi_1 := \sum_{n=0}^{M-1} \left(\phi^{nX,1} + \phi^{a'+nX,r'} + \phi^{a''+nX,r''} \right) + \phi^{a,r}$$

we obtain a structure (v_1, f_1, ϕ_1) on U_1. We see that $\overline{U^{a,r}}$ is located to the left of BOX (as previously, see (3.121)) and so, in the view of (4.47),

$$U_1, U_2 \Subset P \text{ have disjoint closures.} \tag{4.50}$$

Denoting by F^* the total pressure interaction function,

$$F^* := F[v_1, f_1] = \sum_{n=0}^{M-1} \left(F^{nX,1,1} + F^{a'+nX,r',s'} + F^{a''+nX,r'',s''} \right) + F^{a,r,s}$$

$$= H^* + F^{a,r,s},$$

we see that properties (i), (ii) of H^* above (see Step 2) and (4.49) give

$$\begin{cases} F_1^* \ge 0.01 B & \text{in } (-(d-r), d-r) \times (0, \varepsilon r) \supset RECT, \\ F_1^* \ge 8 B & \text{in } SBOX_m, \quad m = 1, \ldots, M, \end{cases} \tag{4.51}$$

cf. (3.118). Moreover

$$v_2 \cdot F^* \ge -1.1 \varepsilon B \quad \text{in } BOX, \tag{4.52}$$

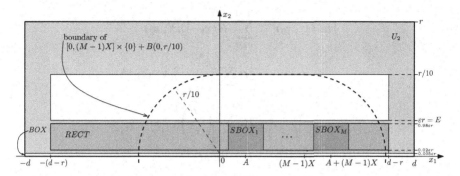

Fig. 4.5 The sets $U_2, BOX, RECT$, and $SBOX_m, m = 1, \ldots, M$ (compare with Fig. 3.13). Note that some proportions are not conserved on this sketch

which is an analogue of the previous relation (3.119) and which we now verify. Let $x \in \operatorname{supp} v_2$ (otherwise the claim is trivial).

Case 1. $x \in [0, (M-1)X] \times \{0\} + B(0, r/10)$. In this case $x_2 \in (0, \varepsilon r)$ (see Fig. 4.5) and

$$-(d-r) < r/10 < x_1 < (M-1)X + r/10 = d/4 + r/10 < d - r,$$

where the leftmost and the rightmost inequalities follow from the fact that $d \geq 10^4 r$ (recall (3.109)). Thus $x \in (-(d-r), d-r) \times (0, \varepsilon r)$ (see Fig. 4.5) and consequently the choice of v_2 (see Lemma 3.11 (iii)) and (4.51) give

$$v_2(x) \cdot F^*(x) = v_{21}(x) F_1^*(x) \geq 0.01 B v_{21}(x) > 0 > -1.1\varepsilon B.$$

Case 2. $x \notin [0, (M-1)X] \times \{0\} + B(0, r/10)$. In this case

$$|H^*(x)| \leq 0.01\varepsilon^2 B, \tag{4.53}$$

which is an analogue of (3.120) and which follows from the decay of H (that is property (iii) in Sect. 3.4.2). Indeed, since in this case

$$|(x_1 - nX, x_2)| \geq r/10, \qquad n = 0, \ldots, M-1,$$

and since $r > 20|A|$ (recall (3.110) we obtain

$$|(x_1 - nX, x_2)| \geq 2|A| \qquad n = 0, \ldots, M-1.$$

Thus we can use property (iii) of H (see Sect. 3.4.2) to write

$$|H^*(x_1, x_2)| \leq \sum_{n=0}^{M-1} |H(x_1 - nX, x_2)|$$

$$\leq 2C \sum_{n=0}^{M-1} |(x_1 - nX, x_2)|^{-4}$$

$$\leq 2C \sum_{n=0}^{M-1} \left(\frac{10}{r}\right)^4$$

$$= 2 \cdot 10^4 C M \varepsilon^4 E^{-4}$$

$$< 0.01 \varepsilon^2 B,$$

where we used (4.46) in the last step. Hence, we obtained (4.53), and so, from our choice of v_2 (namely that $|v_2| \leq 2$, $v_{21} \geq -\varepsilon^2$, $|v_{22}| \leq \varepsilon/2$, recall Lemma 3.11 (iii)) and the bounds on $F^{a,r,s}$ (see (4.49)) we obtain (4.52) by writing

$$v_2(x) \cdot F^*(x) = v_2(x) \cdot H(x) + v_{21}(x) F_1^{a,r,s}(x) + v_{22}(x) F_2^{a,r,s}(x)$$

$$\geq -2(0.01 \varepsilon^2 B) - \varepsilon^2(1.05 B) - \frac{\varepsilon}{2}(0.01 B \varepsilon)$$

$$= -\varepsilon^2 B(0.02 + 1.05 + 0.005)$$

$$\geq -1.1 \varepsilon^2 B.$$

Step 5. We verify (4.3), i.e. that $\{\Gamma_n(G)\}_{n=1}^{M}$ is a family of pairwise disjoint subsets of G_2.

As previously we let

$$z := (A, \varepsilon r/2, 0)$$

and observe that

$$R^{-1}(\Gamma_m(R(BOX))) \subset SBOX_m \qquad m = 1, \ldots, M, \qquad (4.54)$$

which follows in the same way as the previous property (3.114). In fact (3.114) corresponds to the case $m = 1$, and the claim for other values of m follows by translating in the x_1 direction both sides of (3.114) by multiplies of X, see Fig. 4.6. Thus, since the sets $SBOX_m$, $m = 1, \ldots, M$, are pairwise disjoint (recall the comment below (4.48)),

$\{R^{-1}(\Gamma_m(R(BOX)))\}_{m=1}^{M}$ is a family of pairwise disjoint subsets of $RECT$.

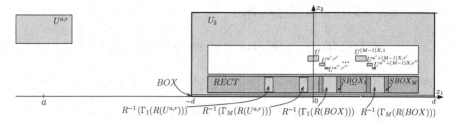

Fig. 4.6 The geometric arrangement for Theorem 4.1 (compare with the previous geometric arrangement, see Fig. 3.14). We note that proportions are not conserved on this sketch

We now show that

$$\left\{ R^{-1}\left(\Gamma_m\left(\overline{U^{a,r}}\right)\right)\right\}_{m=1}^{M} \text{ is a pairwise disjoint family of subsets of } RECT \quad (4.55)$$
$$\text{which are located to the left of } SBOX_1,$$

which is an analogue of the previous relation (3.122), see Fig. 4.6. Here "to the left of" refers to the property that the x_1 coordinate of any point of $R^{-1}\left(\Gamma_M\left(\overline{U^{a,r}}\right)\right)$ is strictly less than the x_1 coordinate of any point of $SBOX_1$; since both $R^{-1}\left(\Gamma_M\left(\overline{U^{a,r}}\right)\right)$ and $SBOX_1$ are rectangles, this is simply

$$\tau(a+r) + A + (M-1)X < A - \kappa E.$$

This inequality can be verified using the facts $\varepsilon < 1/10$ (recall (3.110)) and $\kappa > 1$ (recall (3.109)) by writing

$$\begin{aligned}
\tau(a+r) + (M-1)X &= \tau r(1 - \kappa/\varepsilon) + d/4 \\
&= 0.48r(\varepsilon - \kappa) + \kappa r/4 \\
&= 0.48r\varepsilon - 0.23\kappa r \\
&< 0.48r\varepsilon - 2\varepsilon\kappa r \\
&= \varepsilon r(0.48 - 2\kappa) \\
&< -\kappa\varepsilon r \\
&= -\kappa E,
\end{aligned}$$

as required, where we used the fact that $2\varepsilon < 0.23$ (recall (3.110)) in the first inequality. Property (4.55) is now clear by observing that the sets $R^{-1}\left(\Gamma_m\left(\overline{U^{a,r}}\right)\right)$ are pairwise disjoint (recall each of these sets is a rectangle whose length (in the x_1 direction) is $2\tau r$ (see the comment following (3.122))) and that $X > 2E = 2\varepsilon r > 2\tau r$, where the first inequality holds by our choice of X (recall (4.43)) and the second one is simply that $\tau = 0.48\varepsilon < \varepsilon$) and by recalling the previous property (3.122),

$$R^{-1}\left(\Gamma_1\left(\overline{U^{a,r}}\right)\right) \subset RECT.$$

Properties (4.54) and (4.55) give that

$$\left\{ R^{-1} \left(\Gamma_m(G) \right) \right\}_{m=1}^{M} \text{ is a family of disjoint subsets of } RECT$$

(recall $G = R(\overline{U_1} \cup \overline{U_2})$), which gives (4.3). Indeed, by (4.54) and (4.55),

$$\Gamma_m(G) \subset R(RECT) \subset R(\overline{U_2}) = G_2, \quad m = 1, \ldots, M,$$

and the disjointness follows from the disjointness of the cylindrical projections.

Step 6. Define T, f_2 and ϕ_2 and show the remaining claims (4.11) and (4.12).

(This finishes the construction of the geometric arrangement and thus also finishes the proof of Theorem 4.1.)

Let T, f_2, ϕ_2 be defined as previously (see Sect. 3.4.5). Then (v_2, f_2, ϕ_2) is a structure on U_2 and (4.11) and (4.12) follow in the same way as (3.123), (3.124) in Sect. 3.4.5 by making the following replacements. Replace y by y_n and $SBOX$ by $SBOX_n$ $(n = 1, \ldots, M)$, and use the relations (4.52), (4.54), (4.51), (4.55) instead of the previous relations (3.119), (3.114), (3.118), (3.122) (respectively).

4.6 Blow-Up on a Cantor Set with "Almost Euler Equality", Theorem 1.10

In this section, we prove Theorem 1.10, that is given $\xi \in (0, 1)$, $\vartheta > 0$ we construct a weak solution u to the Navier–Stokes inequality with viscosity $\nu = 0$ (which in this case should perhaps be called "Euler inequality") such that $\xi \le d_H(S) \le 1$, where S is the singular set of u and the "almost equality"

$$-\vartheta \le u \cdot (\partial_t u + (u \cdot \nabla)u + \nabla p) \le 0$$

holds in the sense that

$$-\vartheta \le u \cdot (\partial_t u + (u \cdot \nabla)u + \nabla p) \le 0 \quad \text{everywhere in } \mathbb{R}^3 \times I_k \text{ for every } k, \tag{4.56}$$

for some choice of pairwise disjoint time intervals I_k with $\bigcup \overline{I_k} = [0, \infty)$.

We explain below that u can be obtained by a straightforward sharpening of the construction from Sects. 4.1–4.5, namely, by replacing δ in (4.16) by

$$\delta_j := \min \left\{ \delta, 2\tau^{4j} \vartheta / 3 \right\}. \tag{4.57}$$

We note that such a trick immediately justifies our assumption of zero viscosity. Indeed since our choice of ν_0 in the construction from Sects. 4.1–4.5 is dictated by (3.71) (recall the beginning of the paragraph below (4.17)), we see that (4.57) gives

$$\nu_0 \sup_{x \in R(U_1 \setminus \mathrm{supp}\, \phi_1)} |u[0, f_1](x) \cdot \Delta u[0, f_1](x)| \leq \tau^{4j} \vartheta / 6,$$

and so (by taking $j \to \infty$) we obtain $\nu_0 = 0$.

In order to make the idea of the proof (i.e. that the replacement (4.57) proves Theorem 1.10) more convincing we now briefly go through the main steps of the construction from Sects. 4.1–4.5 articulating the main differences (which are cosmetic) as well as demonstrating how is (4.56) obtained.

Step 1. We construct the geometric arrangement as in Sect. 4.5.

Step 2. For each j We let h_1, h_2 be defined as in (4.16), but with $\delta > 0$ replaced by δ_j, i.e.

$$\begin{aligned}
h_{1,t}^2 &:= f_1^2 - 2t\delta_j \phi_1, \\
h_{2,t}^2 &:= f_2^2 - 2t\delta_j \phi_2 + \int_0^t v_2 \cdot F[v_1, h_{1,r}]\, dr,
\end{aligned} \tag{4.58}$$

and let $h_t := h_{1,t} + h_{2,t}$. Note that, as in Sect. 4.2, $h_1, h_2 \in C^\infty(P \times (-\delta_j, T + \delta_j); [0, \infty))$,

$$(v_i, h_{i,t}, \phi_i) \text{ is a structure on } U_i \qquad \text{for } t \in (-\delta_j, T + \delta_j), i = 1, 2,$$

and

$$h_{2,T}^2(y_n) > \tau^{-2} \left(f_1(R^{-1}x) + f_2(R^{-1}x) \right)^2 + \theta \tag{4.59}$$

for $x \in G, n = 1, \dots, M$, where $y_n = R^{-1}(\Gamma_n(x))$.

Step 3. As in (4.18) We let

$$h_t^{(j)}(x_1, x_2) := \sum_{m \in M(j)} h_t(\pi_m^{-1}(\tau^j x_1), x_2),$$

and let $v^{(j)} \in C^\infty \left(\mathbb{R}^3 \times [0, T]; \mathbb{R}^3 \right)$ be such that conditions (i)–(iv) of Proposition 4.3 are satisfied with $\nu_0 = 0$ and

$$\partial_t |v^{(j)}|^2 + v^{(j)} \cdot \nabla \left(|v^{(j)}|^2 + 2\overline{p}^{(j)} \right) \geq -2\tau^{4j}\vartheta \tag{4.60}$$

in $\mathbb{R}^3 \times [0, T]$ (where $\overline{p}^{(j)}$ is the pressure function corresponding to $v^{(j)}$).

To this end, we repeat the proof of Proposition 4.3 with δ replaced by δ_j and, in order to obtain the extra property (4.60), we amend the calculations from "Case 1" and "Case 2" from Step 3 in Sect. 4.3 as follows. Fix $x \in P$, $t \in [0, T]$ and write, for brevity, $v = v^{(j)}$, $\overline{p} = \overline{p}^{(j)}$.

Case 1. $\phi_1^{\mathfrak{m}}(x) + \phi_2^{\mathfrak{m}}(x) < 1$ for all $\mathfrak{m} \in \{1, \ldots, \mathfrak{M}\}$. Then $v_1^{\mathfrak{m}}(x) = v_2^{\mathfrak{m}}(x) = 0$ and so

$$
\begin{aligned}
\partial_t |v(x, 0, t)|^2 &= -2\delta_j \sum_{\mathfrak{m}=1}^{\mathfrak{M}} (\phi_1^{\mathfrak{m}}(x) + \phi_2^{\mathfrak{m}}(x)) \\
&\geq -2\tau^{4j}\vartheta \\
&= -2\tau^{4j}\vartheta - v(x, 0, t) \cdot \nabla \left(|v(x, 0, t)|^2 + 2\overline{p}(x, 0, t) \right),
\end{aligned}
$$

where we also used (3.44).

Case 2. $\phi_1^{\mathfrak{m}}(x) + \phi_2^{\mathfrak{m}}(x) = 1$ for some $\mathfrak{m} \in \{1, \ldots, \mathfrak{M}\}$. In this case (4.35) (with δ replaced by δ_j) reads

$$
|v_i^{\mathfrak{m}}| \left(\left| \nabla(q_{i,t}^{\mathfrak{m},k})^2 - \nabla(h_{i,t}^{\mathfrak{m}})^2 \right| + 2 \sum_{n=1}^{\mathfrak{M}} \sum_{l=1,2} \left| \nabla p[a_l^{n,k}(t)v_l^n, q_{l,t}^{n,k}] - \nabla p[a_l^{n,k}(t)v_l^n, h_{l,t}^n] \right| \right) \leq \delta_j/2
$$

for $i = 1, 2$, and so

$$
\begin{aligned}
\partial_t |v(x, 0, t)|^2 &= \partial_t q_{1,t}^{\mathfrak{m},k}(x)^2 + \partial_t q_{2,t}^{\mathfrak{m},k}(x)^2 \\
&= -2\delta_j - \left(a_1^{\mathfrak{m},k}(t)v_1^{\mathfrak{m}}(x) + a_2^{\mathfrak{m},k}(t)v_2^{\mathfrak{m}}(x) \right) \cdot \nabla \left(\left(h_{1,t}^{\mathfrak{m}}(x) \right)^2 + \left(h_{2,t}^{\mathfrak{m}}(x) \right)^2 \sum_{n=1}^{\mathfrak{M}} \right. \\
&\qquad \left. + 2 \sum_{n=1}^{\mathfrak{M}} \left(p[a_1^{n,k}(t)v_1^n, h_{1,t}^n](x) + p[a_2^{n,k}(t)v_2^n, h_{2,t}^n](x) \right) \right) \\
&\geq -3\delta_j - \left(a_1^{\mathfrak{m},k}(t)v_1^{\mathfrak{m}}(x) + a_2^{\mathfrak{m},k}(t)v_2^{\mathfrak{m}}(x) \right) \cdot \nabla \left(q_{1,t}^{\mathfrak{m},k}(x)^2 + q_{2,t}^{\mathfrak{m},k}(x)^2 \sum_{n=1}^{\mathfrak{M}} \right. \\
&\qquad \left. + 2 \sum_{n=1}^{\mathfrak{M}} \left(p[a_1^{n,k}(t)v_1^n, q_{1,t}^{n,k}](x) + p[a_2^{n,k}(t)v_2^n, q_{2,t}^{n,k}](x) \right) \right) \\
&= -3\delta_j - v_1(x, 0, t)\partial_{x_1}\left(|v(x, 0, t)|^2 + 2\overline{p}(x, 0, t) \right) \\
&\qquad - v_2(x, 0, t)\partial_{x_2}\left(|v(x, 0, t)|^2 + 2\overline{p}(x, 0, t) \right) \\
&\geq -2\tau^{4j}\vartheta - v(x, 0, t) \cdot \nabla \left(|v(x, 0, t)|^2 + 2\overline{p}(x, 0, t) \right),
\end{aligned}
$$

as required by (4.60), where we used (4.57) and (as in Sect. 4.3) the fact that $\partial_{x_3}|v(x, 0, t)|^2 = \partial_{x_3}\overline{p}(x, 0, t) = 0$.

Step 4. We define the solution $\mathfrak{u}$ and conclude the proof. That is, we let $\mathfrak{u}$ be as in (4.26),

$$\mathfrak{u}(t) := \begin{cases} u^{(j)}(t) & \text{if } t \in [t_j, t_{j+1}) \text{ for some } j \geq 0, \\ 0 & \text{if } t \geq T_0, \end{cases}$$

where

$$u^{(j)}(x_1, x_2, x_3, t) := \tau^{-j} v^{(j)}(\tau^{-j} x_1, \gamma^{-j}(x_2), \tau^{-j} x_3, \tau^{-2j}(t - t_j)),$$

and we let $I_j := (t_j, t_{j+1})$, and $I_\infty := (T_0, \infty)$.

Then (4.60) gives

$$\partial_t \left| u^{(j)} \right|^2 + u^{(j)} \cdot \nabla \left(\left| u^{(j)} \right|^2 + 2p \right) \geq -2\vartheta, \qquad \text{in } \mathbb{R}^3 \times I_j,$$

for every $j \geq 0$, and the rest of the claims of Theorem 1.10 (that is the facts that $\mathfrak{u}$ is a weak solution of the Navier–Stokes inequality (with $\nu_0 = 0$) and that $\xi \leq d_H(S) \leq 1$) follow as in Sect. 4.2.

References

Beale, J. T., Kato, T., & Majda, A. (1984). Remarks on the breakdown of smooth solutions for the 3-D Euler equations. *Communications in Mathematical Physics, 94*, 61–66.

Biryuk, A., Craig, W. & Ibrahim, S. (2007). Construction of suitable weak solutions of the Navier-Stokes equations. In *Stochastic analysis and partial differential equations* (Vol. 429, pp. 1–18). Contemporary Mathematics. Providence, RI: American Mathematical Society.

Blömker, D., & Romito, M. (2009). Regularity and blow up in a surface growth model. *Dynamics of Partial Differential Equations, 6*(3), 227–252.

Blömker, D., & Romito, M. (2012). Local existence and uniqueness in the largest critical space for a surface growth model. *Nonlinear Differential Equations and Applications, 19*(3), 365–381.

Buckmaster, T., & Vicol, V. (2019). Nonuniqueness of weak solutions to the Navier-Stokes equation. *Annals of Mathematics, 189*(1), 101–144.

Caffarelli, L., Kohn, R., & Nirenberg, L. (1982). Partial regularity of suitable weak solutions of the Navier-Stokes equations. *Communications on Pure and Applied Mathematics, 35*(6), 771–831.

Choe, H. J., & Lewis, J. L. (2000). On the singular set in the Navier-Stokes equations. *Journal of Functional Analysis, 175*(2), 348–369.

Constantin, P., & Fefferman, C. (1993). Direction of vorticity and the problem of global regularity for the Navier-Stokes equations. *Indiana University Mathematics Journal, 42*(3), 775–789.

Conte, S. D., & de Boor, C. (1972). *Elementary numerical analysis: An algorithmic approach.* New York-Toronto, Ont.-London: McGraw-Hill Book Co.

Escauriaza, L., Seregin, G. A., & Šverák, V. (2003). $L_{3,\infty}$-solutions of Navier-Stokes equations and backward uniqueness. *Russian Mathematical Surveys, 58*(2), 211–250.

Evans, L. C. (2010). *Partial differential equations* (Second edn., Vol. 19). Graduate Studies in Mathematics. Providence, RI: American Mathematical Society.

Evans, L. C. & Gariepy, R. F. (2015). *Measure theory and fine properties of functions* (Revised edn.). Studies in Advanced Mathematics. Boca Raton, FL: CRC Press.

Falconer, K. (2014). *Fractal geometry–Mathematical foundations and applications* (3rd ed.). Chichester: Wiley.

He, C., Wang, Y., & Zhou, D. (2017). New ε-regularity criteria and application to the box-counting dimension of the singular set in the 3D Navier-Stokes equations. arXiv:1709.01382.

Hopf, E. (1951). Über die Anfangswertaufgabe für die hydrodynamischen Grundgleichungen. *Mathematische Nachrichten, 4*, 213–231. English translation by Andreas Klöckner.

Koh, Y., & Yang, M. (2016). The Minkowski dimension of interior singular points in the incompressible Navier-Stokes equations. *Journal of Differential Equations, 261*(6), 3137–3148.

© Springer Nature Switzerland AG 2019

W. S. Ożański, *The Partial Regularity Theory of Caffarelli, Kohn, and Nirenberg and its Sharpness*, Advances in Mathematical Fluid Mechanics, https://doi.org/10.1007/978-3-030-26661-5

Kukavica, I. (2009a). The fractal dimension of the singular set for solutions of the Navier-Stokes system. *Nonlinearity, 22*(12), 2889–2900.

Kukavica, I. (2009b). Partial regularity results for solutions of the Navier-Stokes system. In *Partial differential equations and fluid mechanics* (Vol. 364, pp. 121–145). London Mathematical Society Lecture Note series. Cambridge: Cambridge University Press.

Kukavica, I., & Pei, Y. (2012). An estimate on the parabolic fractal dimension of the singular set for solutions of the Navier-Stokes system. *Nonlinearity, 25*(9), 2775–2783.

Ladyzhenskaya, O. A., & Seregin, G. A. (1999). On partial regularity of suitable weak solutions to the three-dimensional Navier-Stokes equations. *Journal of Mathematical Fluid Mechanics, 1*(4), 356–387.

Leray, J. (1934). Sur le mouvement d'un liquide visqueux emplissant l'espace. *Acta Mathematica, 63*, 193–248. An English translation due to Robert Terrell is available at http://www.math.cornell.edu/~bterrell/leray.pdf, arXiv:1604.02484.

Lin, F. (1998). A new proof of the Caffarelli-Kohn-Nirenberg theorem. *Communications on Pure and Applied Mathematics, 51*(3), 241–257.

Nečas, J., & Růžička, M., & Šverák, V., (1996). On Leray's self-similar solutions of the Navier-Stokes equations. *Acta Mathematica, 176*(2), 283–294.

Ożański, W. S. (2018). Weak solutions to the Navier–Stokes inequality with arbitrary energy profiles. *Communications in Mathematical Physics* (To appear) preprint available at arXiv:1809.02109.

Ożański, W. S. & Pooley, B. C. (2018). Leray's fundamental work on the Navier-Stokes equations: A modern review of sur le mouvement d'un liquide visqueux emplissant l'espace. In *Partial differential equations in fluid mechanics* (Vol. 452, pp. 113–203). London Mathematical Society Lecture Note series. Cambridge: Cambridge University Press.

Ożański, W. S., & Robinson, J. C. (2019). Partial regularity for a surface growth model. *SIAM Journal on Mathematical Analysis (SIMA), 51*(1), 228–255.

Robinson, J. C. (2011). *Dimensions, embeddings, and attractors* (Vol. 186). Cambridge Tracts in Mathematics. Cambridge: Cambridge University Press.

Robinson, J. C., Rodrigo, J. L. & Sadowski, W. (2016). *The three-dimensional Navier-Stokes equations* (Vol. 157). Cambridge Studies in Advanced Mathematics. Cambridge: Cambridge University Press.

Robinson, J. C., & Sadowski, W. (2009). Almost-everywhere uniqueness of Lagrangian trajectories for suitable weak solutions of the three-dimensional Navier-Stokes equations. *Nonlinearity, 22*(9), 2093–2099.

Scheffer, V. (1976a). Partial regularity of solutions to the Navier-Stokes equations. *Pacific Journal of Mathematics, 66*(2), 535–552.

Scheffer, V., & (1976b). Turbulence and Hausdorff dimension. In *Turbulence and Navier-Stokes equations* (Vol. 565, pp. 174–183). Lecture Notes in Mathematics. Berlin: Springer Verlag. (Proc. Conf., Univ. Paris-Sud, Orsay, 1975)

Scheffer, V. (1977). Hausdorff measure and the Navier-Stokes equations. *Communications in Mathematical Physics, 55*(2), 97–112.

Scheffer, V. (1978). The Navier-Stokes equations in space dimension four. *Communications in Mathematical Physics, 61*(1), 41–68.

Scheffer, V. (1980). The Navier-Stokes equations on a bounded domain. *Communications in Mathematical Physics, 73*(1), 1–42.

Scheffer, V. (1985). A solution to the Navier-Stokes inequality with an internal singularity. *Communications in Mathematical Physics, 101*(1), 47–85.

Scheffer, V. (1987). Nearly one-dimensional singularities of solutions to the Navier-Stokes inequality. *Communications in Mathematical Physics, 110*(4), 525–551.

Seregin, G. (2007). Navier-Stokes equations: almost $L_{3,\infty}$-case. *Journal of Mathematical Fluid Mechanics, 9*(1), 34–43.

Seregin, G. (2012). A certain necessary condition of potential blow up for Navier-Stokes equations. *Communications in Mathematical Physics, 312*(3), 833–845.

Serrin, J. (1963). The initial value problem for the Navier-Stokes equations. In *Nonlinear problems* (pp. 69–98). Madison: University of Wisconsin Press. (Proc. Sympos., Madison, Wis., 1962).

Sohr, H., & von Wahl, W. (1986). On the regularity of the pressure of weak solutions of Navier-Stokes equations. *Archiv der Mathematik (Basel)*, *46*(5), 428–439.

Vasseur, A. F. (2007). A new proof of partial regularity of solutions to Navier-Stokes equations. *Nonlinear Differential Equations and Applications*, *14*(5–6), 753–785.

Index

© Springer Nature Switzerland AG 2019
W. S. Ożański, *The Partial Regularity Theory of Caffarelli, Kohn,
and Nirenberg and its Sharpness*, Advances in Mathematical Fluid Mechanics,
https://doi.org/10.1007/978-3-030-26661-5